Lecture Notes in Computer Science 14153

The series Lecture Notes in Computer Science (LNCS), including its subseries Lecture Notes in Artificial Intelligence (LNAI) and Lecture Notes in Bioinformatics (LNBI), has established itself as a medium for the publication of new developments in computer science and information technology research, teaching, and education.

LNCS enjoys close cooperation with the computer science R & D community, the series counts many renowned academics among its volume editors and paper authors, and collaborates with prestigious societies. Its mission is to serve this international community by providing an invaluable service, mainly focused on the publication of conference and workshop proceedings and postproceedings. LNCS commenced publication in 1973.

Noella Edelmann · Lieselot Danneels ·
Anna-Sophie Novak · Panos Panagiotopoulos ·
Iryna Susha
Editors

Electronic Participation

15th IFIP WG 8.5 International Conference, ePart 2023
Budapest, Hungary, September 5–7, 2023
Proceedings

 Springer

Editors
Noella Edelmann 🆔
University of Continuing Education Krems
Krems an der Donau, Austria

Anna-Sophie Novak 🆔
University of Continuing Education Krems
Krems an der Donau, Austria

Iryna Susha 🆔
Utrecht University
Utrecht, The Netherlands

Lieselot Danneels 🆔
Ghent University
Gent, Belgium

Panos Panagiotopoulos 🆔
Queen Mary University of London
London, UK

ISSN 0302-9743　　　　　　ISSN 1611-3349 (electronic)
Lecture Notes in Computer Science
ISBN 978-3-031-41616-3　　　ISBN 978-3-031-41617-0 (eBook)
https://doi.org/10.1007/978-3-031-41617-0

This Springer imprint is published by the registered company Springer Nature Switzerland AG
The registered company address is: Gewerbestrasse 11, 6330 Cham, Switzerland

Preface

The EGOV-CeDEM-ePart 2023 conference, or short EGOV 2023, is now in the sixth year after the successful merger of three formerly independent conferences, i.e., the IFIP WG 8.5 Electronic Government International Conference (EGOV), the International Conference for E-Democracy and Open Government (CeDEM), and the IFIP WG 8.5 IFIP International Conference on Electronic Participation (ePart). This larger, united conference is dedicated to the broad area of digital or electronic government, open government, smart governance, artificial intelligence, e-democracy, policy informatics, and electronic participation. Scholars from around the world have found this conference to be a premier academic forum with a long tradition along its various branches, which has given the EGOV-CeDEM-ePart conference its reputation as the leading conference worldwide in the research domains of digital/electronic, open, and smart government as well as electronic participation.

The call for papers attracted completed research papers, work-in-progress papers on ongoing research (including doctoral papers), project and case descriptions, as well as workshop and panel proposals. This volume contains full research papers only. All submissions were assessed through a double-blind peer-review process, with at least three reviewers per submission, and the acceptance rate was 36%. The review time took 44 days this year, thanks to the contribution of the many PC members.

The review process was focused on ensuring a double-blind reviewing process and avoiding any conflicts of interest. The track chairs handled the papers within their own track by assigning reviewers and proposing acceptance decisions. The lead track chair became part of the editorial team of the proceedings, in addition to the general chairs. Track chairs were not allowed to submit to their own track, nor were persons from the same university or close collaborators of a track chair allowed to submit, to avoid any conflict of interest. Track chairs assigned the reviewers and selected the program committee members in such a way that there were no conflicts of interest. After at least three reviews were received, the track chairs made a proposal for a decision per paper. The decisions were discussed in a meeting with the general chairs and track chairs to ensure that the decisions were made in a consistent manner across the tracks.

Electronic Government is an evolving field of research and practice. The conference tracks of the 2023 edition reflect the development and progress in this field:

- General E-Government and E-Governance
- General E-Democracy and e-Participation
- ICT and Sustainable Development Goals
- AI, Data Analytics, and Automated Decision Making
- Digital and Social Media
- Digital Society
- Emerging Issues and Innovations
- Legal Informatics

- Open Data: Social and Technical Aspects
- Smart Cities (Government, Districts, Communities and Regions)

38 papers (empirical and conceptual) were accepted for this year's EGOV conference. These LNCS ePart proceedings (LNCS vol. 14153) contain the accepted research papers from the General E-Democracy and e-Participation, ICT & Sustainability, Digital and Social Media, Legal Informatics and Digital Society tracks. The Springer LNCS EGOV proceedings (vol. 14130) contain the papers from the General E-Government and E-Governance; AI, Data Analytics, and Automated Decision Making; Emerging Issues and Innovations; Open Data; and Smart and Digital Cities tracks.

The papers included in these LNCS ePart proceedings (vol. 14153) have been clustered under the following headings:

- e-Participation
- Digital transformation
- Digital technology
- Digital sovereignty

As in the previous years and per the recommendation of the Paper Awards Committee under the leadership of Noella Edelmann (University of Continuing Education Krems, Austria), Evangelos Kalampokis (University of Macedonia, Greece), and Manuel Pedro Rodríguez Bolívar (University of Granada, Spain), the IFIP EGOV-CeDEM-ePart 2023 Conference Organizing Committee granted outstanding paper awards in three distinct categories:

- The most interdisciplinary and innovative research contribution
- The most compelling critical research reflection
- The most promising practical concept

The winners in each category were announced during the awards ceremony at the conference.

The EGOV 2023 conference was hosted by Corvinus University of Budapest. Corvinus University is the best educational institution in Hungary in the fields of economics, management and social sciences. The institution offers state-of-the-art knowledge, a professional network, and a secure future for its 10k+ students. The university has 120+ years of history and 10,000+ students, including 1,500 international students from 80+ nationalities. The institution is ranked in the Top 300 in the QS World rankings for 2021 in the fields of Business and Management, Economics, and Social Sciences. The institution has over 250 partner universities worldwide. It is an AMBA accredited Business institution and the only member of CEMS in Hungary. Corvinus essentially educates the social and economic elite of the region. It strives to produce scientific results that are relevant for Hungary, Europe, and the world. The founders of the university believed that only talent and ambition should count – social or financial status should not prevent anyone from studying. We were very happy to be hosted here and enjoyed the beautiful city of Budapest and the many in-depth discussions advancing the EGOV field.

Many people behind the scenes make large events like this conference happen. We would like to thank the members of the Program Committee, the reviewers, and the track chairs for their great efforts in reviewing the submitted papers. We would also like to express our deep gratitude to Csaba Csáki and his local team at Corvinus University of Budapest for hosting the conference.

We hope that the papers included in this volume will help to advance your research and that you will enjoy reading them.

September 2023

Noella Edelmann
Lieselot Danneels
Anna-Sophie Novak
Panos Panagiotopoulos
Iryna Susha

Organization

Conference Chairs

Csaba Csáki	Corvinus University of Budapest, Hungary
Lieselot Danneels	Ghent University, Belgium
Noella Edelmann	Danube University Krems, Austria
Marijn Janssen	Delft University of Technology, The Netherlands
Evangelos Kalampokis	University of Macedonia, Greece
Ida Lindgren	Linköping University, Sweden
Anna-Sophie Novak	Danube University Krems, Austria
Panos Panagiotopoulos	Queen Mary University of London, UK
Peter Parycek	Fraunhofer FOKUS, Germany/Danube-University Krems, Austria
Gabriela Viale Pereira	Danube University Krems, Austria
Gerhard Schwabe	University of Zurich, Switzerland
Iryna Susha	Utrecht University, The Netherlands
Jolien Ubacht	Delft University of Technology, The Netherlands
Efthimios Tambouris	University of Macedonia, Greece

Program Committee Chairs

Török Bernát	Ludovika – University of Public Service, Hungary
Joep Crompvoets	KU Leuven, Belgium
Csaba Csáki	Corvinus University - Corvinus Business School, Hungary
Lieselot Danneels	Ghent University/Vlerick Business School, Belgium
Noella Edelmann	Danube University Krems, Austria
J. Ramon Gil-Garcia	University at Albany, SUNY, USA
Sara Hofmann	University of Agder, Norway
Marijn Janssen	Delft University of Technology, The Netherlands
Marius Rohde Johannessen	University of South-Eastern Norway, Norway
Evangelos Kalampokis	University of Macedonia, Greece
Hun-Yeong Kwon	Korea University, South Korea
Thomas Lampoltshammer	University for Continuing Education Krems, Austria
Habin Lee	Brunel University, UK

Katarina Lindblad-Gidlund	Mid Sweden University, Sweden
Ida Lindgren	Linköping University, Sweden
Euripidis Loukis	University of the Aegean, Greece
Gianluca Misuraca	Universidad Politécnica de Madrid, Spain
Francesco Mureddu	Lisbon Council, Belgium
Anastasija Nikiforova	Tartu University, Estonia
Anna-Sophie Novak	Danube University Krems, Austria
Panos Panagiotopoulos	Queen Mary University of London, UK
Peter Parycek	Danube-University Krems, Austria
Manuel Pedro Rodríguez Bolívar	University of Granada, Spain
Gerhard Schwabe	University of Zurich, Switzerland
Anthony Simonofskim	Université de Namur, Belgium
Iryna Susha	Utrecht University, The Netherlands
Efthimios Tambouris	University of Macedonia, Greece
Jolien Ubacht	Delft University of Technology, The Netherlands
Gabriela Viale Pereira	Danube University Krems, Austria
Shefali Virkar	Vienna University of Economics and Business, Austria
Anneke Zuiderwijk	Delft University of Technology, The Netherlands

Chair of Outstanding Papers Awards

Noella Edelmann	Danube University Krems, Austria
Evangelos Kalampokis	University of Macedonia, Greece
Manuel Pedro Rodríguez Bolívar	University of Granada, Spain

PhD Colloquium Chairs

Gabriela Viale Pereira	Danube University Krems, Austria
Ida Lindgren	Linköping University, Sweden
J. Ramon Gil-Garcia	University at Albany, SUNY, USA

Web Master

| Gilang Ramadhan | Delft University of Technology, The Netherlands |

Program Committee

Karin Ahlin	Mid Sweden University, Sweden
Suha Alawadhi	Kuwait University, Kuwait
Valerie Albrecht	Danube-University Krems, Austria
Laura Alcaide-Muñoz	University of Granada, Spain
Cristina Alcaide-Muñoz	University of Malaga, Spain
Joao Alvaro Carvalho	University of Minho, Portugal
Renata Araujo	Mackenzie Presbyterian University, Brazil
Wagner Araujo	UNU EGOV, Portugal
Frank Bannister	Trinity College Dublin, Ireland
Ana Alice Baptista	University of Minho, Portugal
Peter Bellström	Karlstad University, Sweden
Flavia Bernardini	Universidade Federal Fluminense, Brazil
Török Bernát	Ludovika – University of Public Service, Hungary
Radomir Bolgov	Saint Petersburg State University, Russia
Alessio Maria Braccini	University of Tuscia, Italy
Paul Brous	Delft University of Technology, The Netherlands
Iván Cantador	Autonomous University of Madrid, Spain
Wichian Chutimaskul	King Mongkut's University of Technology Thonburi, Thailand
Vincenzo Ciancia	Istituto di Scienza e Tecnologie dell'Informazione, Italy
Antoine Clarinval	Université de Namur, Belgium
María Elicia Cortés-Cediel	University Complutense of Madrid, Spain
Joep Crompvoets	KU Leuven, Belgium
Peter Cruickshank	Edinburgh Napier University, UK
Jonathan Crusoe	Gothenburg University and University of Borås, Sweden
Csaba Csaki	Corvinus University of Budapest, Hungary
Frank Danielsen	University of Agder, Norway
Lieselot Danneels	Ghent University, Vlerick Business School, Belgium
Gabriele De Luca	University for Continuing Education Krems, Austria
Bettina Distel	Universität Münster, Germany
Dirk Draheim	Software Competence Center Hagenberg, Austria
Noella Edelmann	Danube University Krems, Austria
Montathar Faraon	Kristianstad University, Sweden
Shahid Farooq	Freelance Governance & Institutional Development Specialist, Pakistan
Cesar Casiano Flores	University of Twente, The Netherlands

Alizée Francey	University of Lausanne, Switzerland
Mary Francoli	Carleton University, Canada
Asbjørn Følstad	SINTEF, Norway
Jonas Gamalielsson	University of Skövde, Sweden
Francisco García Morán	European Commission, Luxembourg
Luz-Maria Garcia	Universidad de la Sierra Sur, Mexico
Mila Gasco-Hernandez	University at Albany, SUNY, USA
Alexandros Gerontas	University of Macedonia, Greece
Sarah Giest	Leiden University, The Netherlands
J. Ramon Gil-Garcia	University at Albany, SUNY, USA
Dimitris Gouscos	University of Athens, Greece
Malin Granath	Linköping University, Sweden
Annika Hasselblad	Mid Sweden University, Sweden
Marcus Heidlund	Mid Sweden University, Sweden
Moreen Heine	Universität zu Lübeck, Germany
Marissa Hoekstra	TNO, The Netherlands
Sara Hofmann	University of Agder, Norway
Hanne Höglund Rydén	University of Agder, Norway
Roumiana Ilieva	Technical University of Sofia, Bulgaria
Tomasz Janowski	Gdańsk University of Technology, Poland
Marijn Janssen	Delft University of Technology, The Netherlands
Björn Johansson	Linköping University, Sweden
Moon-Ho Joo	Personal Information Protection Commission, Japan
Yury Kabanov	National Research University Higher School of Economics, Russia
Natalia Kadenko	Delft University of Technology, The Netherlands
Evangelos Kalampokis	University of Macedonia, Greece
Nikos Karacapilidis	University of Patras, Greece
Areti Karamanou	University of Macedonia, Greece
Naci Karkin	Pamukkale University, Türkiye
Ilka Kawashita	University of Phoenix, USA
Jong Woo Kim	Hanyang University, South Korea
Fabian Kirstein	Fraunhofer FOKUS, Germany
Ralf Klischewski	German University in Cairo, Egypt
Michael Koddebusch	University of Münster, ERCIS, Germany
Peter Kuhn	fortiss, Germany
Hun-Yeong Kwon	Korea University, South Korea
Thomas Lampoltshammer	University for Continuing Education Krems, Austria
Habin Lee	Brunel University London, UK
Hong Joo Lee	The Catholic University of Korea, South Korea

Azi Lev-On	Ariel University Center, Israel
Katarina Lindblad-Gidlund	Mid Sweden University, Sweden
Ida Lindgren	Linköping University, Sweden
Mikael Lindquist	University of Gothenburg, Sweden
Johan Linåker	RISE Research Institutes of Sweden, Sweden
Martin Lněnička	University of Pardubice, Czechia
Euripidis Loukis	University of the Aegean, Greece
Rui Pedro Lourenço	INESC Coimbra/FEUC, Portugal
Edimara Luciano	Pontifical Catholic University of Rio Grande do Sul, Brazil
Luis F. Luna-Reyes	University at Albany, SUNY, USA
Bjorn Lundell	University of Skövde, Sweden
Johan Magnusson	University of Gothenburg/Kristiania University College, Sweden
Stanislav Mahula	KU Leuven, Belgium
Heidi Maurer	University for Continuing Education Krems, Austria
Keegan McBride	University of Oxford, UK
John McNutt	University of Delaware, USA
Ulf Melin	Linköping University, Sweden
Sehl Mellouli	Université Laval, Canada
Tobias Mettler	University of Lausanne, Switzerland
Morten Meyerhoff Nielsen	United Nations University EGOV, Portugal
Gianluca Misuraca	Universidad Politécnica de Madrid, Spain
Solange Mukamurenzi	University of Rwanda, Rwanda
Francesco Mureddu	Lisbon Council, Belgium
Anastasija Nikiforova	University of Tartu, Estonia
Anna-Sophie Novak	Danube University Krems, Austria
Panos Panagiotopoulos	Queen Mary University of London, UK
Peter Parycek	Danube-University Krems, Austria
Luiz Paulo Carvalho	Federal University of Rio de Janeiro, Brazil
Laura Piscicelli	Utrecht University, The Netherlands
Nina Rizun	Gdańsk University of Technology, Poland
Manuel Pedro Rodríguez Bolívar	University of Granada, Spain
Marius Rohde Johannessen	University of South-Eastern Norway, Norway
Alexander Ronzhyn	University of Koblenz-Landau, Germany
Boriana Rukanova	Delft University of Technology, the Netherlands
Michael Räckers	University of Münster, ERCIS, Germany
Rodrigo Sandoval-Almazan	Universidad Autónoma del Estado de Mexico, Mexico
Verena Schmid	Donau-Universität Krems, Austria
Hendrik Scholta	University of Münster, ERCIS, Germany

Gerhard Schwabe	University of Zurich, Switzerland
Walter Seböck	University for Continuing Education Krems, Austria
Andreiwid Sheffer Corrêa	Federal Institute of Sao Paulo, Brazil
Tobias Siebenlist	Rhine-Waal University of Applied Sciences, Germany
Kerley Silva	University of Porto, Portugal
Anthony Simonofski	Université de Namur, Belgium
Leif Sundberg	Mid Sweden University, Sweden
Iryna Susha	Utrecht University, The Netherlands
Øystein Sæbø	University of Agder, Norway
Efthimios Tambouris	University of Macedonia, Greece
Luca Tangi	Joint Research Centre - European Commission, Italy
Lörinc Thurnay	Danube University Krems, Austria
Daniel Toll	Linköping University, Sweden
Jolien Ubacht	Delft University of Technology, the Netherlands
Marco Velicogna	IRSIG-CNR, Italy
Gabriela Viale Pereira	Danube University Krems, Austria
Shefali Virkar	Vienna University of Economics and Business, Austria
Gianluigi Viscusi	Linköping University, Sweden
Jörn von Lucke	Zeppelin Universität Friedrichshafen, Germany
Bianca Wentzel	Fraunhofer FOKUS, Germany
Guilherme Wiedenhöft	Federal University of Rio Grande, Brazil
Elin Wihlborg	Linköping University, Sweden
Mete Yildiz	Hacettepe Üniversitesi İİBF, Türkiye
Maija Ylinen	Tampere University of Technology, Finland
Sang Pil Yoon	Korea University, South Korea
Chien-Chih Yu	National Chengchi University, Taiwan
Thomas Zefferer	Graz University of Technology, Austria
Dimitris Zeginis	University of Macedonia, Greece
Anneke Zuiderwijk	Delft University of Technology, the Netherlands

Additional Reviewers

Jörg Becker	University of Münster, ERCIS, Germany
Bettina Distel	University of Münster, ERCIS, Germany
Corinna Funke	public GmbH, Germany
Amirhossein Gharaie	Linköping University, Sweden
Junchul Kim	Brunel University London, UK

Ini Kong	Delft University of Technology, the Netherlands
Yannik Landeck	fortiss, Germany
Changwon Park	Brunel University London, UK
Elham Shafiei Gol	Brunel University London, UK
Yao Hua Tan	Delft University of Technology, The Netherlands

Contents

Digital Sovereignty

E-Participation

The Case for a Broader Approach to e-Participation Research: Hybridity, Isolation and System Orientation

Martin Karlsson[1]([⊠]) [iD] and Magnus Adenskog[2] [iD]

[1] Örebro University, Örebro, Sweden
martin.karlsson@oru.se
[2] Lund University (Affiliated Researcher, CIRCLE), Helsingborg, Sweden
magnus.adenskog@ch.lu.se

Abstract. Two decades into the young history of e-participation research, we aim to take stock of the state of this field in the light of three developments that we argue have substantial implications for research on electronic participation: (1) dissolving boundaries between online and offline spheres of political participation; (2) academic isolation of e-participation research from other research fields related to political participation; and (3) the systemic turn in research on political participation. In relation to these developments, we discuss the potential role of the field in the future and make the case for a broader approach to e-participation research.

Keywords: E-participation · Hybridization · Academic isolation · Political systems · Democratic innovations

1 Introduction

At the start of the new millennium "electronic participation" or "e-participation" gained increasing attention within government as a concept delineating processes of citizen participation in politics aided by or administered through ICTs [1, 2]. As the internet and ICTs in general diffused across the developed world, visions for how these technologies could aid and even revolutionize democratic practices [cf. 3] materialized in the form of processes that took advantage of novel ICTs to aid citizens' participation in politics. Some of these simply transferred "offline" models for political participation into the digital sphere, while others created new forms of political participation [4].

In concurrence with the increased utilization of ICT-enabled or aided processes of citizen participation, a new academic field arose related to the concept of e-participation. In its infancy, the field of e-participation research was viewed as a sub-field of "electronic democracy" or "e-democracy" [5], which was understood as a wider field encompassing

This paper is dedicated to the memory of Professor Joachim Åström 1973–2022. This research was supported by the research foundation Formas (grant number: 2021-00089).

N. Edelmann et al. (Eds.): ePart 2023, LNCS 14153, pp. 3–14, 2023.
https://doi.org/10.1007/978-3-031-41617-0_1

questions related to how "ICT-supported communication processes can facilitate democratic goals" [6, p. 373]. However, the popularity of e-participation rose steadily in the 2000s and soon became the more widely used concept in academic research (see Fig. 1 below).

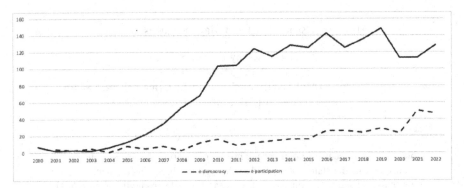

Fig. 1. Number of articles with e-democracy and e-participation in title, as author keyword or in abstract, 2000–2022 (Scopus).

While there are various definitions of e-participation available in the literature, one of the most often cited definitions comes from Macintosh [43] who argues that e-participation is focused on the use of ICTs for online dialogue, deliberation and consultation between citizens and government. In this paper we will focus specifically on e-participation in the context of political decision making. We will for instance therefore not consider e-participation in the realm of e-service development. Further, we will focus specifically on "invited spaces" for e-participation [7], that is, government-initiated processes of political participation, rather than bottom-up forms of citizen participation (e.g. social movements, protests, activism).

Lindner and colleges [8] underscore three pivotal factors for the onset of e-participation practices at the turn of the millennium. (1) *A crisis of democratic legitimacy* that gave rise to a discourse highlighting the need for democratic renewal [see also 9]. This discourse can be seen as a window of opportunity for democratic renewal and experimentation with new democratic practices. (2) *Technological affordances of new ICTs* offering unprecedented possibilities for effective and interactive communication. (3) *New normative ideals for democratic government* as deliberative and participatory democracy gained broad support not only in academia but also in governmental institutions [10, 11].

Two decades into the young history of e-participation research, we aim to take stock of this field in the light of three developments that we argue have substantial implications for research on electronic participation: (1) dissolving boundaries between online and offline spheres of political participation; (2) academic isolation of e-participation research from other research fields related to political participation; and (3) the systemic turn in research on political participation. In relation to these developments, we discuss the potential role of e-participation research in contemporary societies. What new research questions arise? What theoretical and methodological development is warranted?

2 Dissolving Boundaries and Hybridization

The pace of technological development within the area of ICTs is matched by the speed by which these technologies inhibit more and more aspects of our lives and societies [12]. One central dimension of this societal immersion in ICTs is the blurring of the lines between the online and offline spheres, not least facilitated by the development and diffusion of mobile technologies. Diamankati [13] has defined our current relationship to ICTs as a "post-desktop paradigm" characterized by a detachment of the internet from place. As we no longer access the internet from a computer statically located at a definite place, but as De Souza E Silva and Sheller [14, p. 4] write, "carry it with us", our transports between online and offline spheres are more frequent and less noticeable. According to Šimůnková [15, p. 49], this has blurred and undermined distinctions such as "[a]bsence/presence, here/there, close/far, public/private, real/virtual". At its essence, this relationship with technology presents a state of hybridity, as clear distinctions between online and offline are not only becoming harder to make but also less valuable.

2.1 The Hybridization of Politics

These changes are also obvious in the political sphere. Today, information about political processes and developments, political debate and discourse, as well as channels for political influence, are primarily found online or in hybrid settings [11, 15]. The state of hybridity in politics been most authoritatively defined by Chadwick [16], who investigates how political actors function within an environment that is hybridized between new and old, online and offline and tailor their repertoires of action based on this hybridity. For instance, Chadwick and others [e.g. 17] have studied the repertoires of action of what they call "new hybrid mobilization movements". These political movements utilize new as well as old media logics to effectively mobilize supporters and influence policy-making. New media (meaning ICTs in general and social media in particular) is utilized to monitor the views of their member base and coordinate action. However, offline political protests or manifestations are often the forms of political action preferred by these movements, and old media is the target of these actions [16]. Other movements, such as the "Fridays for future" climate movement, organize localized offline political actions, not least "climate strikes", and utilize social media to boost the impact of such actions [18].

There are also indications of a hybridization of invited spaces for political participation online. This trend is illustrated through an analysis of cases in the Participedia database [19] (in Fig. 2 below). Participedia consists of global reports on processes of political participation. While the database consists of both invited spaces for participation as well as bottom-up organized participatory processes, there is a clear skewness towards the former. In Fig. 2, the number of cases in the database with instances of online participation is plotted by year from 2000 to 2022. The number of cases in the database has grown intensely over the last two decades. However, the growth is disproportionately leaning towards hybrid participation cases, meaning combinations of face-to-face and ICT-enabled participation. At the same time, the number of participation cases exclusively facilitated online has been relatively stable. Seemingly, hybridity has increasingly become the norm in e-participation, according to the database. The only exceptions to this rule are the years most clearly affected by the Covid-19 pandemic and the lockdown policies that accompanied the pandemic in many countries, which meant that participatory processes had to go online.

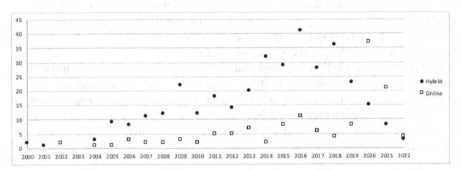

Fig. 2. Number of cases of citizen participation characterized as "Hybrid" and "Online" in the Participedia database 2000–2022.

To the extent that the cases reported in the Participedia database are representative of the implementation of e-participation processes in the world, this trend has strong implications for e-participation research. It indicates that the dissolving boundaries between online and offline are also evident within the field of invited spaces for political participation. This development immediately raises questions: To what extent has e-participation research adapted to this changing reality? To what extent does this research field engage with theories of hybridity and empirical cases of hybrid participation?

Misinikov and colleagues [20] suggest that the e-participation field has engaged with these aspects to a limited extent; they argue that "e-participation scholarship lacks sufficient conceptual consolidation to reflect upon the fundamental changes in digital technology that occurred over the past decade or so". There are, however, notable exceptions in individual research contributions analyzing hybrid cases of e-participation and engaging with questions related to this hybridization [cf. 21, 22]. Further, the hybridity concept is present in e-participation research [cf. 23], although with a different meaning. In this context, hybridization is used to connote e-participation processes that combine web 1.0 and web 2.0 technology.

Hybridization presents an important role for e-participation research. Farrell argues that paradoxically, the increasing integration of ICTs into all aspects of political interactions will lead to fewer rather than more political scientists specializing in the internet [24]. As the intersections between the internet and politics become more plentiful and diverse, this relationship becomes the business of all political scientists rather than a specialized sub-field. However, according to Farrell, hybridization requires more rather than less specialization. As the internet becomes "both ubiquitous and invisible," its intermediating role risks being taken for granted [24, p. 47]. He argues that political science is in need of "unbundling the Internet into discrete (yet sometimes mutually reinforcing or undermining) mechanisms" [24, p. 47]. This call for unbundling the internet can be seen as naive given the pace and diversity of technological development and utilization; however, it could potentially point to an important focus area for e-participation research. For the broader fields of research focused on political participation to fully understand participation in contemporary societies, there is a great need for better theorization of the mechanisms related to technology that affects participation.

3 Academic Isolation

Academic isolation is one potential risk of organizing research on the intersection of political participation and ICTs in a distinct research field (e-participation) with field-specific concepts, publication outlets and conferences. Academic isolation can be defined as a state of a research field characterized by relative disconnection to adjacent research fields that share commonalities in terms of themes, research objects, theories, and methodologies. Isolation is problematic for at least two reasons: (1) isolation can mean that the field takes fewer research perspectives into account in theorizing and empirically studying its object of research (influence from), and (2) it can also mean that the research in the field has less influence on other adjacent research fields (influence on) [25, p. 1672]. Thus, academic isolation may be detrimental to knowledge production within the field as well as its impact in other fields of research.

Academic isolation may be especially detrimental for academic fields that produce knowledge about the intersection between fields of knowledge. Drawing inspiration from and producing knowledge relevant to adjacent fields is essential for such intersecting fields of research. This can be argued to be the case for e-participation research that is not only a multidisciplinary field of research but also a field that addresses a thematic area at the intersection between information technology and political participation.

We will consider the level of isolation of e-participation research from other research fields related to political participation. The degree of academic isolation of e-participation research is measured through a bibliometric network analysis of research publications using the network analysis software VOSviewer [26]. The analysis focuses on cross-citation (citing other publications within the sample) and co-citation (citing the same references as other publications within the sample) between research publications in research fields related to political participation. The sample of publications analyzed is the 1632 most-cited English language publications in the Web of Science database with author keywords including e-participation, democratic innovations, deliberative democracy and political participation. This list of keywords is not comprehensive but chosen to reflect central concepts within the field as well as relatively new developments within the research field (democratic innovations and e-participation). The network visualization (Fig. 3 below) indicates $\geq$ 10 cross- or co-citations as a tie between publications; it represents the number of citations of a publication as the size of a node. In total, 345 publications had 10 or more cross- or co-citations with other publications and were thus included in the network map. Clusters of publications were created based on the smart local moving algorithm [27] with a threshold number for clusters of a minimum of 20 nodes (i.e. publications).

Four clusters were identified, which, based on our review (focused on the most central publications within each cluster), are labelled: *e-participation* (red), *communication studies* (green), *political participation* (yellow) and *deliberative democracy and democratic innovations* (blue).

The e-participation cluster is most isolated from the other clusters, sharing the fewest co- and cross-citations with publications in the other clusters. There are articles in the e-participation cluster, not least some of the most well-cited articles [e.g. 28], that share connections to articles in two of the other clusters (green and blue). Overall, however, the analysis indicates that the e-participation literature is largely disconnected from the research literature in adjacent fields within research on political participation. This is true to a lesser extent for the other clusters, as the number of cross and co-citations between publications in these clusters are magnitudes greater.

The furthest distance between clusters in the analysis is identified between the clusters named e-participation (red) and political participation (yellow). The political participation cluster consists largely of seminal works within political science, that develop and evaluate theories explaining variations in citizens' participation in politics [e.g., 29]. While such central nodes in the political participation cluster share strong connections to other clusters, they are largely absent in the e-participation cluster.

Given that such publications precede the formation of the e-participation field, such disconnection could be interpreted as representing what has been termed above as a lack of "influence from" such research. In other words, e-participation research, to a small extent, has been influenced by central works within the field of research on political participation.

Turning to the "influence on" side of the coin, to what extent does e-participation research have an influence on other fields of research related to political participation? According to this analysis, it is hard to find instances of "influence on", meaning that few publications in the e-participation cluster have 10 or more connections to publications in other clusters they precede (are published before). Here, we should remember that the bar set for connections within the network map is quite high (at 10 or more co- and cross-citations). However, it cannot be seen as a good sign for the influence of e-participation research that few candidates for cross-cluster influential studies emerge from the analysis.

This analysis indicates the academic isolation of e-participation research in relation to other fields of research related to political participation. Such isolation may be detrimental to knowledge production as well as the impact of e-participation research. However, the network analysis presented above gives only a superficial picture of the connectedness of e-participation research to adjacent fields based solely on co- and cross-citation. There is a need of more research investigating the transfer of theories and concepts between these adjacent field, for instance through systematic literature reviews.

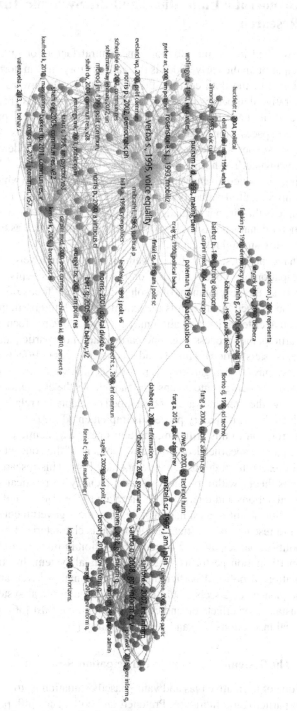

Fig. 3. Cross and co-citations between top cited articles using the keywords: E-participation, democratic innovations, deliberative democracy, deliberation, political participation, citizen participation and participatory democracy. (Color figure online)

4   Democratic Innovations and the Systemic Turn in Participation Research

As described in the introduction, one central factor for understanding the rise of e-participation in the early 2000s is the participatory shift in normative democratic theory at the end of the 20th century. From the late 1960s onward, participatory, direct and deliberative democracy arose as central normative ideals for democratic government. These ideals have heavily influenced research on political participation, not least on "invited spaces" for participation, often conceptualized as "democratic innovations" [30]. These ideals have also had a great influence on how democratic innovations have been evaluated. The normative democratic theories have created yardsticks for evaluation leading to a value-driven evaluation focused on the extent to which a participatory process lives up to central values within a specific normative theory of democracy [31]. One example is the discourse quality index, used to evaluate the extent to which deliberative democratic innovations live up to deliberative values such as justification, universalism, constructivism and respect [32].

In recent years, value-driven evaluation and normatively driven research on political participation have been critiqued on two essential accounts. First, value-driven evaluation risks the development of a solely micro-level focus of evaluation. As the central question of such evaluation is the extent to which the participatory process lives up to normative criteria, its evaluation may be biased towards focusing on internal aspects of the participatory process (e.g. who participates, how participants communicate, and participants' satisfaction). Thus, broader consequences or effects of such processes (macro aspects) may be disregarded [31, p. 46]. Second, value-driven evaluation has been criticized for not being context-sensitive enough. As the evaluation criteria are set by general normative theories, they are not developed or adapted in relation to the contextual setting in which the participatory process is implemented [33].

In relation to this criticism, there has been a systemic turn in research on political participation, spearheaded by the development of the concept of "deliberative systems" within research on deliberative democracy [34]. This research direction falls back on systems theory within political science, identifying political systems as the sum of all political actions and interactions that relate to the policy- and decision-making process in a political unit (e.g., a nation-state, or a local government) [35]. The systemic approach to research on political participation is characterized by a functional perspective on political participation. The central question in this research is what functions political participation performs within the political system. In this strand of research, the evaluation of political participation is functionality-driven and hence, focused on the consequences or effects of participation on the political system. These outcomes can, for instance, be effects on and changes in political trust [36], political knowledge [37], political institutions [38] and decision quality [39].

4.1   The Systemic Turn and e-Participation Research

Critique of normative bias and value-based evaluation is, to some extent, echoed within the e-participation literature. Pratchett and colleagues [40, p. 190] argue that "[m]uch of the literature focuses on exploring particular normative accounts of deliberative or representative forms of democracy". Grönlund [5, p. 13] argues that e-participation research rests on the assumption that "direct democracy is the ideal value for eParticipation" and that e-participation processes may lead societies towards direct democracy. However, theoretical and methodological tools to transgress this normative orientation have not been developed within e-participation research. There are, however, constructive contributions that share elements of the system-oriented research on democratic innovations.

For instance, Kubicek and Aichholzer [41] argue for a "relativity theory" for evaluating e-participation processes, meaning that criteria and methods of evaluation are tailored to the type of e-participation process evaluated, rather than striving for a unifying, one-size-fits-all, evaluation framework. This constitutes a step in the right direction, as it facilitates an adaptation of evaluation frameworks to the character of the e-participation process. However, the systemic perspective offers a second important insight that the characteristic of the political system in which e-participation processes are implemented must be taken into account to understand what systemic functions this process can and does play. One example of such an analysis is offered by Åström and colleges [42]. Through a comparative analysis of e-participation processes in Sweden, Estonia and Iceland, they illustrate that institutional and circumstantial factors in political systems strongly influence the role and impact of e-participation processes.

We argue that e-participation research could benefit from a "systemic turn", characterized by a greater focus on macro aspects of e-participation processes and a functionality-driven evaluation. Such a direction of e-participation research could be a way to overcome the critique of normative bias and facilitate a better understanding of the functions e-participation performs in political systems. A first step would be to connect the research fields by harvesting the knowledge produced in the research fields visible in Fig. 3.

5   Concluding Discussion

In this paper, we have discussed the state of e-participation research in relation to three developments with important implications for the field. In this short conclusion, we aim to sketch out suggestions for future directions of e-participation research in relation to these developments.

The dissolving boundaries between online and offline spheres imply that the scope of political practices and events relevant to e-participation research may be broadened. ICTs hold a central or complementary role in many (if not most) forms of political participation today. Therefore, the knowledge and expertise within e-participation research are arguably applicable and valuable in relation to a wide variety of participatory practices. Further, as Farrell [24] argues, the immersion of politics on the internet may make technological aspects of political participation less noticeable or be taken for granted by researchers. Hence, specialists in the area of e-participation may have much to contribute to the understanding of contemporary forms of participation in various stages of hybridization between online and offline.

However, indications of academic isolation of e-participation research suggest that specialized knowledge from e-participation researchers is not transferred to adjacent fields to any substantial extent. Isolation, however, goes both ways. The lack of connection between democratic innovation- and e-participation research also indicates that Farrell might have been right in predicting that political scientists disregard the importance of unbundling the mechanisms of the internet and thus do not seek to draw lessons from the e-participation field.

We see several benefits of strengthening the connection between these fields. As stated above, such connections could strengthen the understanding of technological aspects of political participation in an age of technological hybridization. Further, e-participation research could draw inspiration from the systemic turn in research on political participation, which creates an avenue for grappling with the issues of normative bias in e-participation research that have received criticism [5, 6, 40]. There are some studies that have started to investigate e-participation in similar ways, e.g. Wirtz et al. [44], but this approach needs to broaden. The systemic turn in general and functionality-driven evaluation in particular are directions that may advance the field towards a greater understanding of macro-level aspects of e-participation and more context-sensitive research of e-participation processes.

All in all, we have made the case for a broader approach to e-participation research. We argue that the field should broaden its empirical focus to include the variety of participatory practices that have been technologized in this era of hybridization. Further, the e-participation field should be more open to adjacent research fields related to political participation. Lastly, the field should broaden its theoretical and methodological scope to better encompass macro aspects and systemic functions of e-participation processes.

References

1. OECD: Citizens as Partners: Information, Consultation and Public Participation in Policy-making. OECD Publishing (2001)
2. OECD: Promises and Problems of e-Democracy; Challenges of Citizen On-line Engagement. OECD Publishing (2003)
3. London, S.: Teledemocracy vs. deliberative democracy: a comparative look at two models of public talk. J. Interpersonal Comput. Technol. 3(2), 33–55 (1995)
4. Oser, J., Hooghe, M., Marien, S.: Is online participation distinct from offline participation? A latent class analysis of participation types and their stratification. Polit. Res. Q. 66(1), 91–101 (2013)
5. Grönlund, Å.: ICT is not participation is not democracy–eParticipation development models revisited. In: Macintosh, A., Tambouris, E. (eds.) ePart 2009, LNCS, vol. 5694, pp. 12–23. Springer, Heidelberg (2009). https://doi.org/10.1007/978-3-642-03781-8_2
6. Susha, I., Grönlund, Å.: eParticipation research: systematizing the field. Gov. Inf. Q. 29(3), 373–382 (2012)
7. Kersting, N.: Online participation: from 'invited' to 'invented' spaces. Int. J. Electron. Gov. 6(4), 270–280 (2013)
8. Lindner, R., Aichholzer, G., Hennen, L.: Electronic Democracy in Europe. Prospects and Challenges of e-Publics, e-Participation and e-Voting. Springer, Cham (2016). https://doi.org/10.1007/978-3-319-27419-5
9. Norris, P. (ed.): Critical Citizens: Global Support for Democratic Government. Oxford University Press, Oxford (1999)
10. Amnå, E.: Playing with fire? Swedish mobilization for participatory democracy. J. Eur. Publ. Policy 13(4), 587–606 (2006)

11. Karlsson, M.: Digital democracy and the European Union. In: Bakardjieva Engelbrekt, A., Leijon, K., Michalski, A., Oxelheim, L. (eds.), The European Union and the Technology Shift, pp. 237–261. Springer, Cham (2021). https://doi.org/10.1007/978-3-030-63672-2_10

12. Graham, M., Dutton, W.H.: Society and the Internet: How Networks of Information and Communication are Changing Our Lives, 2nd edn. Oxford University Press, Oxford (2019)

13. Diamantaki, K.: The ambiguous construction of place and space. In: Gehman, U., Reiche, M. (eds.) Real Virtuality: About the Destruction and Multiplication of World, pp. 245–268. Transcript-Verlag, Bielefeld (2014)

14. De Souza E Silva, A., Sheller, M. (eds.) Mobility and Locative Media. Mobile Communication in Hybrid Spaces. Routledge, New York (2014)

15. Šimůnková, K.: Being hybrid: a conceptual update of consumer self and consumption due to online/offline hybridity. J. Mark. Manag. 35(1–2), 40–74 (2019)

16. Chadwick, A.: The Hybrid Media System: Politics and Power, 2nd edn. Oxford University Press, Oxford (2017)

17. Karpf, D.: The MoveOn Effect: the Unexpected Transformation of American Political Advocacy. Oxford University Press, Oxford (2012)

18. Boulianne, S., Lalancette, M., Ilkiw, D.: "School strike 4 climate": social media and the international youth protest on climate change. Media Commun. 8(2), 208–218 (2020)

19. Fung, A., Warren, M.E.: The participedia project: an introduction. Int. Publ. Manag. J. 14(3), 341–362 (2011)

20. Misnikov, Y., Filatova, O., Trutnev, D.: Empirical modeling of e-participation services as media ecosystems. In: Meiselwitz, G. (ed.) HCII 2021. LNCS, vol. 12774, pp. 87–104. Springer, Cham (2021). https://doi.org/10.1007/978-3-030-77626-8_6

21. Åström, J., Grönlund, Å.: Online consultations in local government: what works, when, and why. connecting democracy: online consultation and the flow of political communication. In: Coleman, S., Shane, P. (eds.) Connecting Democracy, pp. 75–96. MIT Press, Cambridge (2012)

22. Thoneick, R.: Integrating online and onsite participation in urban planning: assessment of a digital participation system. Int. J. E-Plann.Res. 10(1), 1–20 (2021)

23. Charalabidis, Y., Loukis, E.: Participative public policy making through multiple social media platforms utilization. Int. J. Electron. Gov. Res. 8(3), 78–97 (2012)

24. Farrell, H.: The consequences of the internet for politics. Annu. Rev. Polit. Sci. 2012(15), 35–52 (2012)

25. Chi, P.-S., Conix, S.: Measuring the isolation of research topics in philosophy. Scientometrics 127(4), 1669–1696 (2022). https://doi.org/10.1007/s11192-022-04276-y

26. Van Eck, N., Waltman, L.: Software survey: VOSviewer, a computer program for bibliometric mapping. Scientometrics 84(2), 523–538 (2010)

27. Waltman, L., van Eck, N.J.: A smart local moving algorithm for large-scale modularity-based community detection. Eur. Phys. J. B 86(11), 1–14 (2013). https://doi.org/10.1140/epjb/e2013-40829-0

28. Sæbø, Ø., Rose, J., Flak, L.S.: The shape of eParticipation: characterizing an emerging research area. Gov. Inf. Q. 25(3), 400–428 (2008)

29. Verba, S., Schlozman, K.L., Brady, H.E.: Voice and Equality: Civic Voluntarism in American Politics. Harvard University Press, Cambridge (1995)

30. Smith, G.: Democratic Innovations: Designing Institutions for Citizen Participation. Cambridge University Press, Cambridge (2009)

31. Adenskog, M.: Democratic Innovations in Political Systems: Towards a Systemic Approach. Örebro university, Örebro (2018)

32. Steenbergen, M.R., Bächtiger, A., Spörndli, M., Steiner, J.: Measuring political deliberation: a discourse quality index. Comp. Eur. Polit. 1, 21–48 (2003)

33. Mansbridge, J., et al.: A systemic approach to deliberative democracy. In: Parkinson, J., Mansbridge, J. (eds.) Deliberative Systems. Cambridge University Press, New York (2012)
34. Parkinson, J., Mansbridge, J. (eds.) Deliberative Systems: Deliberative Democracy at the Large Scale. Cambridge University Press, Cambridge (2012)
35. Easton, D.: An approach to the analysis of political systems. World Polit. **9**(3), 383–400 (1957)
36. Åström, J., Jonsson, M.E., Karlsson, M.: Democratic innovations: reinforcing or changing perceptions of trust? Int. J. Publ. Adm. **40**(7), 575–587 (2017)
37. Grönlund, K., Setälä, M., Herne, K.: Deliberation and civic virtue: lessons from a citizens deliberation experiment. Eur. Polit. Sci. Rev. **2**(1), 95–117 (2010)
38. Adenskog, M.: After the equilibrium: democratic innovations and long-term institutional development in the city of Reykjavik. Anal. Kritik **40**(1), 31–54 (2018)
39. Drazkiewicz, A., Challies, E., Newig, J.: Public participation and local environmental planning: Testing factors influencing decision quality and implementation in four case studies from Germany. Land Use Policy **46**, 211–222 (2015)
40. Pratchett, L., Durose, C., Lowndes, V., Smith, G., Stoker, G., Wales, C.: Empowering communities to influence local decision making: systematic review of the evidence (2009)
41. Kubicek, H., Aichholzer, G.: Closing the evaluation gap in e-participation research and practice. IN: Evaluating e-Participation: Frameworks, Practice, Evidence, pp. 11–45 (2016)
42. Åström, J., Hinsberg, H., Jonsson, M.E., Karlsson, M.: Crisis, innovation and e-participation: towards a framework for comparative research. In: Wimmer, M.A., Tambouris, E., Macintosh, A. (eds.) ePart 2013. LNCS, vol. 8075, pp. 26–36. Springer, Heidelberg (2013). https://doi.org/10.1007/978-3-642-40346-0_3
43. Macintosh, A.: Characterizing e-participation in policy-making. In: Proceedings of the 37th Annual Hawaii International Conference on System Sciences, pp. 117–126 (2004)
44. Wirtz, B.W., Daiser, P., Binkowska, B.: E-participation: a strategic framework. Int. J. Publ. Adm. **41**(1), 1–12 (2018)

Identifying Institutional, Contextual and Dimension-Based Patterns in Public Strategic Planning Processes

Cristina Alcaide Muñoz[1]([✉]), Manuel Pedro Rodríguez Bolívar[2],
Laura Alcaide Muñoz[2], and Miguel Morales Marín[2]

[1] Department of Accounting and Finance, University of Malaga, Málaga, Spain
`c.alcaide@uma.es`
[2] Department of Accounting and Finance, University of Granada, Granada, Spain

Abstract. To enhance the quality of life of citizens and face urban challenges, governments increasingly pursue the successful development and implementation of smart initiatives based on the intensive use of information and communication technologies (ICTs). The institutional environment and the context where public policies are adopted are key factors for their successful development and implementation. This study seeks to contribute prior research analysing whether patterns about the type, collaborative aspects and strategic planning approaches can be identified in SCs analysing the institutional environment, the context, and the smart dimensions in which these processes are immersed. To achieve this aim, this paper analyses 1,635 SC strategy planning approaches undertaken by 12 Spanish SCs using the institutional context and the smart dimensions as key enablers for these patterns. Findings reveals that there is significant influence of institutional, context and dimension-based smart initiatives on the strategic planning patterns.

Keywords: Smart Cities · Strategic Planning · Institutional context · Smart Dimensions

1 Introduction

Increasing urban population have arisen new challenges in the urban space for which cities were not prepared. To face these new urban challenges successfully, Poister (2010) [1] already predicted that strategic planning would play increasingly a key role in public sector entities in the 2020s decade due to the need of anticipating and managing change adroitly and effectively. Indeed, Johnsen (2022) [2] has recently indicated that public managers perceive the net benefit of the strategic planning as positive, especially to anticipate response for both turbulent environments (international conflicts, financial crises, etc.) and new urban challenges.

In this regard, strategic planning processes have demonstrated to provide benefits in terms of managing the interactions with stakeholders [3, 4], being considered relevant as a governance tool especially in the urban areas where local governments implemented the

© IFIP International Federation for Information Processing 2023
Published by Springer Nature Switzerland AG 2023
N. Edelmann et al. (Eds.): ePart 2023, LNCS 14153, pp. 15–32, 2023.
https://doi.org/10.1007/978-3-031-41617-0_2

Information and Communication Technologies (ICTs) to reform city governance under the the Smart City (SCs) wave [5]. Under this framework, new governance models are proposed based on actively involving and collaborating with stakeholders [6] - 'marriage' between technology and urban governance [5]-, seeking to introduce participative and collaborative governance models.

Nonetheless, although the implementation of ICTs into the city governments has been considered itself a strategy for administrative reform in many countries [7], the SC trend has often lacked an integrated strategic urban plan for a comprehensive SC program [8]. Indeed, Dameri & Benevolo (2016) [9] found that no strategic plan explicitly includes SCs among the strategies for urban planning and development pursued in an administrative cycle. This has made that the way SCs have been developed across countries in the world has been highly diverse, mainly due to their individual visions and priorities for achieving their specific goals [10].

In fact, considering that the use of long-term strategic visions and strategic plans are at the heart of public governance [11], the results obtained from strategic planning processes have been different [12] and the relationship between strategic planning and both managerial and citizen perceptions of performance has been found dependent on improved external relations [4]. This could explain why in the first stages of government digitalization processes not visible impacts in the short term were [13].

In addition, although strategic planning in SC enhances the efficient performance of government tasks, allowing the successful achievement of desired goals [14], the way strategies are formulated and adopted varies significantly across city governments (formal/informal, vertical/transversal, and so on) [15]. In this regard, although Mintzberg (1994) [16] was critical with the formal strategic planning processes (due to their focus on the private sector), strategic planning (when formal and comprehensive) seems to have a positive impact on both organizational performance and on measuring an organization's ability to achieve its goals [17], even although its results could be often just modestly satisfactory [18].

In any case, strategic planning is highly contingent upon its context, including the characteristics of public sector organizations, the institutional environment, the type of client served by public sector organizations and whether strategic planning is linked to broader strategy implementation activities [19]. Indeed, prior research has demonstrated that the managerial institutional factor [20] and the political context [21] have a significant effect on strategic planning for goal achievement. Therefore, to achieve positive outcomes, a strong understanding of contextual conditions, governance models, and public value is needed to both understand and develop realistic smart city strategies [5].

Recent prior research has demonstrated that demographical city profile, citizen profile living in the SCs, and political factors could help to explain the design of strategic planning processes into the SC framework [15, 21, 22]. Also, the most widely accepted approach to a SC includes six smart dimensions that play a fundamental role in the design of the SC strategy [23].

These factors (socio-economic attributes and smart dimensions) get into the institutional theory domain and, concretely, at the coercive level, focused on socio-economic political pressures [24] as ones of more implicated factors into the institutional isomorphism defined by DiMaggio & Powell (2000) [25]. Indeed, this theory can help us to understand the way that institutions respond to the environmental pressures and cultural expectations under uncertain situations in a uniform way over time, using mimetic mechanisms increasing homogenization in their actions [26]. Besides, the institutional isomorphism may help examine the diffusion of similar organizational strategies and structures for assessing the influence of elite interests [25].

Nonetheless, as far as we are concerned, there is a lack of studies analysing if homogenous patterns on strategic planning processes -a standard way of performing strategic planning- in SCs can be identified by both the institutional context and the smart dimension domains. Therefore, this paper seeks to contribute prior research analysing whether patterns about the type, collaborative aspects and strategic planning approaches can be identified in SCs analysing the institutional environment, the context and the smart dimensions in which these processes are immersed. Concretely, the research questions to be analysed in this study are: RQ1. are both the institutional factors and the context of SCs determinant on strategic planning patterns concerning Smart Initiatives? and RQ2. do the Smart Dimensions have influence on how to strategic planning patterns concerning Smart Initiatives?. To achieve this aim, this paper analyses 1,635 SC strategy planning approaches undertaken by 12 Spanish SCs using the institutional context and the smart dimensions as key enablers for these patterns [19, 27].

Although smart initiatives are implemented in different levels of public administration [23], special attention should be paid to the local government level due to its role in the construction of urban infrastructures, the implementation of local policies and investment of financial resources [28].

The remainder of this paper is as follows. Next section describes the research and data collection methods. Then we show the findings for each one of the research questions posed in this paper. Finally, the conclusions and discussions section bring the paper to an end.

2   Research Methodology and Sample Selection

2.1   Sample Selection and Variables Used

Spain is a noteworthy European country when it comes to the implementation of smart initiatives [15, 29]. It presents an attractive subject for investigation, with the aim of improving comprehension regarding the formulation, implementation, and societal impact of such initiatives. This study represents an initial step towards exploring the profile of the Spanish government, particularly with respect to the management and communication of strategic information relating to smart projects. The findings provide valuable empirical evidence for other countries in their strategic planning for similar initiatives.

The process of data collection occurred in two phases. The first phase involved identifying Spanish cities labelled as "Smart City" in two well-known rankings: the European project sponsored by Asset One Inmobilienentwicklungs AG (accessible at http://www.smart-cities.eu), and the EUROCITIES network (available at http://www.eurocities.eu). Based on a thorough review of the information provided, the authors selected twelve

Spanish SCs as a sample for their study. The focus of this study is on large-scale Spanish cities at the local level, with populations ranging from 200,000 to 4 million inhabitants. This selection was made because these cities face a multitude of challenges in planning, designing, financing, constructing, governing, and operating urban infrastructure and services. The classification and factors reflected in these European SCs rankings [30] were also taken into consideration during the selection process.

After the Spanish cities were identified, the authors established criteria for determining the strategy documents of these cities based on the framework proposed by Yigitcanlar (2018) [31]. This marked the beginning of the second phase, which was divided into two sub-phases. Firstly, between January and February 2021, the authors obtained access to the strategies developed by the cities under study via their official websites. Secondly, these strategies were analysed in detail, resulting in a total of 1,635 cases of smart initiatives. These initiatives were categorized according to the department responsible for the project in the smart city, the smart city domain, stakeholders involved, vision, objectives, and strategies for the city's smart transformation.

As for the variables analysed, this study focuses on analysing the key attributes that differentiate the various patterns of organizing the strategic planning of smart initiatives in a SC [15]. In addition to analysing the demographic attributes of the city and its citizens to identify patterns in strategic planning for smart initiatives (see Table 1), since they have a significant impact in discovering patterns in e-government policies and online information disclosure [32].

2.2   Method Used

To accomplish the aim of this study, a two-step data analysis approach was employed, first a cluster analysis is carried out and then difference tests. The first step involved conducting a hierarchical cluster analysis to group municipalities with institutional and contextual factors common, thereby achieving a characterization of similar municipalities. The Ward method (also known as minimum inertia loss method) [33] was specifically utilized, which links cases to minimize variance within each group. To measure the strength of the cluster analysis, the principal component analysis was developed, and the Kaiser-Meyer-Olkin (KMO) measure of sampling adequacy is 0.893 with sig. $< .001$, revealing a proper cluster analysis. The outcomes of the cluster analysis are presented in Table 2.

During the second phase, we aimed to identify the differences between clusters based on the formalization of smart strategies, public-private collaboration, and the strategy approach (top-down or bottom-up). Also, we identified the differences between clusters focused on smart dimensions, and if these dimensions give rise to differences between the formalization of smart strategies, public-private collaboration, and the strategy approach. To carry out this analysis, we first carry out the Kolmogorov-Smirnov normality test, which show that the considered variables followed a no normal distribution. Therefore, we carried out Kruskal Wallis different tests to identify the above-mentioned differences.

Table 1. Definition of variables used in this study.

Attribute	Definition	Calculation
Type of Strategic Planning[a]	Approach used for strategic planning into the Smart City	0 = Informal 1 = Formal
Collaboration[a]	Responsible body of the smart initiative	0 = No collaboration 1 = Public-private collaboration
Strategic Planning Approach[a]	Strategic planning approach when a Smart City initiative is implemented	0 = Top-Down 1 = Bottom-Up
Political ideology[b]	Political ideology of the ruling party	0 = Progressive 1 = Conservative
Political Strength[c]	Numerical variable that reflects the local governments' level of political strength (in percentage)	$\sum_{i=0}^{n} s_i^2 / s^2$ Where: S = Total councillors in municipality $S\,i$ = Councillors in political party "i"
Political Stability[b]	The time that the same political party has remained in power, despite having held the municipal elections	Years in the power
Population[b]	Population residing in the municipality	Number of inhabitants
Population density[b]	The measurement of population per unit area	Population/Km2
Age of Inhabitants[b]	Age of inhabitants	Age 15 from 24 Age 25 from 34 Age 35 from 64
Level of Education[b]	Level of inhabitants with secondary education	Inhabitants with secondary education
	Level of inhabitants with superior education	Inhabitants with superior education
Income per capita[b]	Income per capita	Income (thousand euros) per capita

Notes: [a]Local Government Website, [b]National Statistical Institute (INE) (www.ine.es/) and [c]Herfindahl index is used
Source: Own Elaboration

3   Analysis of Results

3.1   RQ1 – Are both the Institutional Factors and the Context of SCs Determinant on Strategic Planning Patterns Concerning Smart Initiatives?

The Table 2 shows the statistical description of institutional and context factors under study in each cluster. We can observe that smart cities with the largest population is in cluster 3, whose numbers of inhabitants are between 1,620,809 and 3.182.981, and whose mean value of population density numbers 10,878.97. Additionally, the inhabitants of these cities are highly qualified (superior education-30.5-mean value), and the most of them are in the 35–64 age range.

The maximum level of income per capita is in cluster 1 (30,889 €); however, there is a broad spread of data, because the income per capita in Bilbao is much higher than Malaga and Murcia. In fact, Bilbao is considered the second Spanish cities with the highest per capita income in Spain (the first one is San Sebastian-Donostia) due to being an important port and industrial city. Also, the Guggenheim Museum and the rehabilitation of the statuary environment placed the city on the world tourist map. Similarly, there is a large dispersion of data related to superior education, since in cities such as Bilbao, a large number of inhabitants have superior education, nearly 35%. However, in Malaga and Murcia, only 20% of their population have a higher education degree.

As for institutional factors, we can observe that SCs of cluster 1 and 3 are led by conservative and progressive parties, respectively. In cluster 1, there is greater political strength, since its mean values is nearly 40; however, the other ones have a similar political strength (mean value around 30). Finally, in cluster 1, we can find the SCs with greater political stability (15 years), by contrast, in cluster 3 are those where political parties only led the municipality during a municipal legislature (3 years).

Kruskal Wallis' test shows significant difference among cluster and strategic patterns under study, particularly, formalization and public-private collaboration (90% confidence level). However, there does not appear to be any empirical difference between the strategic planning approach when a smart initiative is adopted in cities (top-down/bottom-up) and SCs of each cluster. In this sense, based on the information disclosed by SCs, most of them in cluster 1 and 2 formulate smart initiatives in detail, above 80%. However, SCs in cluster 3 opt to develop informal strategies, it can be caused by the short time of parties leading cities, hindering long-term policy development.

Moreover, even though public-private collaboration is not predominant in Spanish smart initiatives, Cluster 2 is the focal point for encouraging collaboration between public and private entities, given that 8% of smart initiatives are jointly developed with private entities (Table 3).

3.2   RQ2 – Do the Smart Dimensions have Influence on Strategic Planning Patterns Concerning Smart Initiatives?

The Table 4 shows smart dimensions promoted by SCs in each cluster. For instance, SCs mainly favour social cohesion, cultural services, healthy environment and safety and tourist attraction (smart living). If we focus on tourism sector, Spain is the main European Country that receive the most international tourists. In fact, according to National

Table 2. Descriptive statistics of variables used per cluster.

	CLUSTER 1: Bilbao, Malaga and Murcia					CLUSTER 2: Valencia, Sevilla, A Coruña, Zaragoza, Gijón, Valladolid and Terrasa				
	Mean	Median	Std. Dev.	Min	Max	Mean	Median	Std. Dev	Min	Max
Population	452,452.67	443,243.00	112,229.71	345,110.00	569,002.00	453,541.00	299,715.00	247,872.39	216,428.00	787,808.00
Population density	3,426.11	1,440.04	4,279.69	500.30	8,338.00	3,410.76	3,079.07	2,299.79	682.84	6,452.52
Political ideology	1.00	1.00	-	1.00	1.00	0.14	0.00	-	0.00	1.00
Political Strength	39.96	40.00	2.91	37.03	42.85	29.02	26.92	3.86	25.00	34.62
Political Stability	15.00	15.00	0.00	15.00	15.00	3.57	3.00	1.51	3.00	7.00
Age 15-24	9.85	10.29	1.42	8.27	11.00	8.83	9.18	1.04	7.22	9.94
Age 24-35	12.07	12.63	1.14	10.76	12.81	11.00	11.08	1.05	9.58	12.12
Age 35-64	44.34	44.13	0.67	43.80	45.09	44.80	44.43	0.95	44.05	46.77
Secondary Educ.	10.36	10.21	0.39	10.06	10.79	11.06	11.12	0.77	10.26	12.47
Superior Educ.	24.96	10.61	8.20	19.85	34.42	25.03	24.85	2.53	20.63	28.88
Income per capita	24,000.67	20,688.00	5,966.92	20,425.00	30,889.00	22,071.29	22,273.00	548.26	21,301.00	22,922.00

(continued)

Table 2. (*continued*)

		CLUSTER 3: Madrid and Barcelona			
	Mean	Median	Std. Dev	Min	Max
Population	2,401,895.00	2,401,895.00	1,104,622.42	1,620,809.00	3,182,981.00
Population density	10,878.97	10,878.97	7,954.29	5,254.44	16,503.50
Political ideology	0.00	0.00	-	0.00	0.00
Political Strength	30.04	30.04	7.13	25.00	35.08
Political Stability	3.00	3.00	0.00	3.00	3.00
Age 15-24	8.99	8.99	0.06	8.95	9.03
Age 24-35	13.57	13.57	0.78	13.02	14.12
Age 35-64	43.32	43.32	0.80	42.75	43.88
Secondary Educ.	13.44	13.44	3.51	10.96	15.92
Superior Educ.	30.50	20.50	3.48	28.04	32.96
Income per capita	28,217.00	28,217.00	931.97	27,558.00	28,876.00

Source: Own Elaboration

Table 3. Difference between the main strategic patterns in each cluster

Kruskal Wallis' test	Formalization	Public-private Collaboration	Approach
Cluster 1 – Cluster 2	−41.09*	−29.41**	−3.50
Cluster 1 – Cluster 3	−689.13***	−18.35	−9.41
Cluster 2 – Cluster 3	−648.04***	11.05	−5.92

Source: Own Elaboration * $p < 0.1$; ** $p < 0.05$; *** $p < 0.01$

Statistical Institute (INE), in January 2023, more than four million people have visited Spain to enjoy its clime, culture and gastronomy.

Furthermore, a very interesting mix of culture coexists in Spain, with more than forty ethnic groups, because of its history and the geographical position. According to Spanish ministry of foreign affairs and the European Commission [34, 35], Spain is one of the three main European countries (along with Germany and Italy) with the highest immigration flow. All this encourages the need of social policies aimed at promoting social cohesion, education and so on [36]. Therefore, it makes sense that SCs put their efforts and resources in tackling these challenges.

The second most promoted smart dimension by SCs of clusters 1 and 3, is smart governance (19.93% and 25.79%, respectively). However, in cluster 2, SCs choose to develop initiative related to environment issues (21.24%), which is the thirst smart dimension fostered by SCs in cluster 1, along with smart economics policies (18.45% in both dimension).

In cluster 3, SCs pursue the development of innovation and entrepreneurship, given that almost 25% of smart initiatives adopted for them favour the increased local business fabric (smart economy). By contract, these cities do not pay special attention to mobility issues. Perhaps, this dimension does not need to be developed in Madrid and Barcelona, because they, given their characteristics (i.e. their large population and population density), promoted it in the past and, currently have the necessary infrastructures.

Table 4. Proportion of Smart Initiatives by Smart Dimensions in each cluster

	CLUSTER 1	CLUSTER 2	CLUSTER 3
Mobility	17.34%	16.85%	5.03%
Economy	18.45%	17.68%	24.53%
People	2.21%	6.64%	12.58%
Living	23.62%	23.90%	26.42%
Governance	19.93%	13.69%	25.79%
Environment	18.45%	21.24%	5.66%
Total Smart Initiatives per cluster	**16.57%**	**73.70%**	**9.72%**

Source: Own Elaboration

We can observe in Table 5, the difference between smart dimensions, both individually and each cluster, and the three types of strategic patterns under study, that is, formal strategic planning, public-private collaboration and the strategic planning approach when a smart initiative is adopted in smart cities (top-down/bottom-up).

Analysing smart dimensions individually, regarding the formalization of strategy, there is significant difference between both smart environment and smart living and, governance, economy, and mobility. Likewise, we can observe significant difference between smart economy and all other dimensions as well as between smart people and economy. In this sense, the development of formal strategy is most likely in those smart initiatives that address mobility (87.55%), qualification of human and participation in public life (smart people – 78.88%) and social cohesion, healthy environment, and cultural services (smart living – 77.88), followed by transparency of governance systems, participation in decision-making and availability of public services (smart governance – 64.61%).

Regarding public-private collaboration, we identify significant difference between smart economy and, mobility, people, living, governance, and environment. It is also found between smart mobility and, people and living as well as between smart living and environment. Although the public-private is poorly promoted by local governments, it is mainly developed in smart initiatives related to innovation and entrepreneurship (13.58%). By contrast, when smart initiatives focus on qualification of human, participation in public life, cultural services, social cohesion, healthy environment and tourist attractions, the public and private collaboration is poor or non-existent (below 4%).

In general, the predominant approach is top-down, that is, local governments are the leaders of smart initiatives. In this respect, special attention may be made of smart initiative aimed at the improvement of ecosystem, where none of them has been led by citizenry; however, citizens are involved in smart initiatives concerning business fabric, entrepreneurship, and business innovation (almost 20%).

Paying attention to each cluster, we identify no significant difference between smart dimension related to the strategic planning approach when a smart initiative is adopted in smart cities, in cluster 1. Notwithstanding, there is significant difference in terms of the formalization of strategy and the collaboration between public and private entities.

Concerning the formalization of strategy, in cluster 1, Table 5 shows significant difference between smart mobility and, environment, living and governance. SCs of cluster 1 mainly formulate formal smart strategies on environmental issues (92%) and transparency policies and the citizenry involvement in decision-making process, and the accessibility to public services (89.06%), followed closely by smart initiatives on social cohesion, healthy environment, tourist attraction and cultural services (88.9%). In respect of public and private collaboration, there is also significant difference between smart economy and, mobility, economy, living and governance. When smart initiatives promote economics and people issues, the collaboration between public and private entities is more likely (16 and 14.3%, respectively). However, this collaboration is absent in topics such as the accessibility of public services and environmental improvement.

In cluster 2, significant difference between dimensions concerning the three strategic patterns under study is identified. In this sense, there is significant difference between all smart dimensions, except for smart people and smart mobility as well as smart governance and smart environment. The development of formal strategy is most likely in those smart initiatives aimed at the improvement of qualification of human and participation in public life and, in turn, mobility (97.5 and 94.81%, respectively), followed by smart initiatives on social issues (88.15%).

Regarding public and private collaboration, in cluster 2, there is significant difference between smart people and, mobility, economy, governance and environment. It is also found between smart economy and, mobility, governance, and environment. Likewise, the Table 5 shows significant difference between smart living and, mobility and environment. In this sense, SCs of cluster 2 mainly promote public and private collaboration on economic and mobility issues (14.08 and 10.34%, respectively). Additionally, significant difference between smart dimensions, in term of the strategic planning approach when a smart initiative is adopted in cities, is identified, only between smart environment and, government and economy, and the latter and mobility.

Finally, cluster 3 also shows significant difference between dimensions related to the three strategic patterns being analysis. Concerning the formalization of strategy, there is significant difference between both smart mobility and environment, living, governance and economy, since the formal strategic planning is only developed in mobility smart initiatives. Just a significant difference is identified in term of public and private collaboration between smart people and governance, given that private entities are involved in 15% of initiatives aimed at the participation of citizens in public life and their qualification. In addition, regarding strategic approach, we can observe significant difference between smart people and, living, environment, governance, economy, and mobility, since citizens mainly take an active part in smart initiative on their education and qualification and the participation in public life.

4   Discussions

Findings reveal three different patterns in strategic planning of smart initiatives. Whereas cluster 1 is characterised by SCs with a differentiated political environment from those of the other clusters (conservative political parties, high level of political strength and greater political stability), SCs in cluster 3 are mainly characterised by a high urban population with higher educational level and a higher income per capita (see Table 2). In addition, SCs in cluster 3 are led by progressive parties.

Adding to the different political setting, the main differences between clusters 1 and 2 are focused on the more homogeneous and higher level of population and higher population density in cluster 1 vs the more homogeneous income per capita and higher education level of urban residents in cluster 2. In brief, our research identifies three different clusters according to the institutional and contextual environment of sample SCs.

This different institutional and contextual environment make us to identify different patterns in strategic planning processes. According to the findings of our research, whereas SCs in cluster 3 are focused on informal strategic planning processes, SCs in clusters 1 and 2 use formal strategic planning processes. This finding confirms prior research that indicate that conservative parties are more likely to design formalized

Table 5. Difference between Smart Dimensions, including each cluster.

Kruskal Wallis' test	Formalization	Public-private Collaboration	Approach	CLUSTER 1 Formalization	Public-private Collaboration	Approach
Environment – Living	-29.76	-36.79**	10.39	3.98	2.12	-2.12
Environment – Governance	86.79***	-25.09	9.43	4.22	0.00	0.00
Environment – Economy	152.09***	51.26***	16.23**	8.13	21.68***	0.00
Environment – People	-19.05	-29.11	22.91**	27.87	19.36*	0.00
Environment – Mobility	-100.58***	13.46	6.36	30.40***	2.95	2.95
Living – Governance	-116.55***	-11.70	0.97	-0.24	2.12	-2.12
Living – Economy	181.85***	88.10***	5.84	4.15	19.56***	-2.12
Living – People	10.71	7.67	12.51	23.89	17.24	-2.12
Living – Mobility	-70.82*	50.25***	-4.04	26.42***	0.83	0.83
Governance – Economy	65.30**	76.35***	6.81	3.92	21.68***	0.00
Governance – People	-105.84**	-4.02	13.48	23.66	19.36*	0.00
Governance – Mobility	-187.37***	38.55**	-3.07	26.18***	2.95	2.95
Economy – People	171.14***	80.38***	-6.68	-19.74	2.32	0.00
Economy – Mobility	-252.67***	-37.80**	-9.87	22.27**	-18.73***	2.95
People – Mobility	-81.53*	42.58*	-16.55	2.53	-16.41	2.95

Table 5. (continued)

Kruskal Wallis' test	CLUSTER 2			CLUSTER 3		
	Formalization	Public-private Collaboration	Approach	Formalization	Public-private Collaboration	Approach
Environment – Living	-81.53***	-37.31***	8.39	0.00	3.79	0.00
Environment – Governance	11.33	-17.60	10.95*	0.00	1.94	0.00
Environment – Economy	87.38***	30.70**	14.13**	0.00	6.12	2.04
Environment – People	-137.80***	-54.09***	7.53	0.00	11.93	7.95*
Environment – Mobility	-120.23***	8.19	2.97	-9.94***	9.94	0.00
Living – Governance	-92.87***	-19.70	-2.56	0.00	1.85	0.00
Living – Economy	168.92***	68.01***	5.74	0.00	2.33	2.04
Living – People	-56.27*	-16.78	-0.87	0.00	8.14	7.95***
Living – Mobility	-38.70*	45.50***	-5.43	-9.94***	6.15	0.00
Governance – Economy	76.05***	48.30***	3.19	0.00	4.18	2.04
Governance – People	-149.13***	-36.49*	-3420.00	0.00	0.99*	7.95***
Governance – Mobility	-131.56***	25.79	-7.98	-9.94***	7.99	0.00
Economy – People	225.19***	84.79***	6.61	0.00	-5.81	-5.91**
Economy – Mobility	-207.61***	-22.51	-11.17*	-9.94***	3.82	-2.04
People – Mobility	17.57	62.28***	-4.56	-9.94***	-1.99	-7.95*

Source: Own Elaboration * $p < 0.1$; ** $p < 0.05$; *** $p < 0.01$

smart strategies than progressive political parties [35, 37] and are less likely to directly promote citizen participation [38].

In addition, this difference could also be due to the political stability aspect, which is shorter and, especially, on the demographic profile in which both are immersed. By contrast, no differences between strategic planning processes in SCs in cluster 3 vs SCs in clusters 1 and 2 exist concerning the collaboration and approach aspects. In short, findings seem to indicate that demographic profiles of SCs could influence on the formalization of the strategic planning processes, but not on the collaboration or on the approach used.

Also, considering demographic aspects as main enablers for formalization aspects in strategic planning processes in SCs, the citizens' educational level and income per capita seem to be those attributes that most influence on the collaboration aspect in strategic planning processes. This finding seems to indicate that collaboration with private companies is produced when these attributes (educational level and income per capita) are homogenous in the population and are higher. That's why there is not significant difference in these aspects between SCs in cluster 2 and 3. This is a novel finding of this research that would be of merit for future deeper research investigation.

In brief, findings seem to suggest that the institutional and contextual environment are relevant to design strategic planning processes in SCs, which confirms the institutional isomorphism concept (coercive isomorphism) included into the institutional theory [26], which stems from the political influence and institutional legitimacy [25]. In fact, this research has found different patterns of strategic planning processes according to the institutional and contextual environment of the analysed smart initiatives.

By contrast, findings do not show differences in strategic planning processes concerning the approach. Therefore, future research could make deeper analysis in SCs of different countries to identify other attributes, such as the administrative culture of the countries, that could help to explain differences in strategic planning processes in these SCs.

Concerning the dimension aspect of smart initiatives, findings suggest that smart initiatives undertaken are focused on the different urban challenges according to the context of the different cities. Indeed, although initiatives in smart living (social cohesion, cultural services, healthy environment and safety and tourist attraction) seem to be a shared concern among the different clusters, smart governance initiatives are promoted in SCs in clusters 1 and 3, whereas environment issues are mainly faced in SCs in cluster 2.

Perhaps, this finding could be due to the different demographical aspects of the SCs, mainly in the heterogeneity/homogeneity aspect of the population density. Cluster 1 and 3 show heterogeneously low and high population density, respectively, which makes them not to be concerned with the environment issues (in the first case, due not to be a challenge yet and, in the second one, because they have previously implemented initiatives concerning environment issues which makes them to think have control over this aspect). SCs in cluster 2 are homogeneous concerning the problem of population density and they think that the environment concern is a main urban challenge to be solved at this stage of their population development.

Future research directions should focus on grouping SCs according to the population density to deeper analyse whether this aspect is influencing of eroding the environment in the urban area and examine how these SCs are dealing with this problem. Also, future studies using similar development stages of SCs in different countries could contribute to know whether these urban policies to solve environmental challenges is a shared feeling in cities with homogenous and high population density on each one of the development stages of SCs, with the aim at getting insights concerning the link between smart initiatives and development stages in SCs.

Focusing on the smart dimension to which the smart initiatives are addressed, findings suggest significant differences mainly on both strategic planning formalization and the collaboration aspect among the different smart dimensions, especially between smart environment and smart living vs the rest of smart dimensions. Only significant differences are found between smart environment and both smart economy and people in the approach aspect. Nonetheless, these differences could not be longer required because prior research indicates the need of a systemic and integrated approach that enhances the interoperability and scalability of solutions [39].

In any case, this research has identified that formal strategic planning processes are mainly present in smart mobility, people and living, whereas collaboration with the private sector is mainly present in smart initiatives related to innovation and entrepreneurship. By contrast, smart initiatives addressed to obtain social impact, the collaboration is almost inexistent. This finding could be due to the benefit-oriented goal of private companies, which seems not to be interested in supporting aspects concerning social impact on the urban areas.

In this regard, public administrations should establish public policies for promoting private companies' involvement in improving the social aspect of the urban area in which they operate as a part of their sustainability aspect. In this regard, future research directions should deeper analyse the factors and drivers that could explain the findings of this research as well as how to involve all stakeholders in improving the social aspects of the urban residents leaving economic benefits ahead.

Finally, significant differences among formalization, collaboration and approach aspects into the strategic planning processes exist when analysing the smart dimensions in the different clusters previously defined. Whereas in SCs in cluster 2 significant differences are usual in the formalization and collaboration aspects in all smart dimensions, these differences are focused on smart mobility vs other dimensions (mainly, environment, living and governance) in SCs in cluster 1 and 3. Approach differences are mainly present in SCs in cluster 3 when comparing smart people and other dimensions (environment, living, governance and economy). These differences are mainly identified due to the different profile of the clusters previously mentioned (mainly on the demographic and the citizen profile aspects).

5 Conclusions

This research sought to identify whether the institutional, contextual, and dimension-based attributes could define different patterns about the type, collaborative aspects and strategic planning approaches in the smart initiatives developed by a sample of Spanish SC.

Using cluster analysis, this research has identified three different clusters mainly characterised by different political setting and different demographical profile (population and population density). These different political and demographical settings influences on different patterns in strategic planning processes of smart initiatives, especially those linked to the type of strategic planning and the collaboration approaches. Also, the smart dimension to which the smart initiatives are addressed have influence on the strategic planning patterns in both as separate aspect and as an aspect jointly with the cluster analysis made.

Therefore, this paper has confirmed the institutional isomorphism concept included into the institutional theory [26], examining different strategic planning patterns in SCs according to the institutional, contextual, and dimension-based aspects. Nonetheless, this research has found that the approach does not seem to be different into SCs belonging into one country. All this opens new avenues for future research, including the analysis of enablers, drivers and inhibitors for different patterns in strategic planning processes as well as the need of comparative analysis in SCs in different countries to know if aspects like the administrative culture, the political culture, the individualism aspect, or the power distance in the country [40] can have an impact on the strategic planning processes in SCs.

Acknowledgments. The authors disclosed receipt of the following financial support for the research, authorship, and/or publication of this article: This research was funded by financial support from Regional Government of Andalusia, Spain (research project numbers P20_00314 and B-SEJ-556-UGR20).

References

1. Poister, T.H.: The future of strategic planning in the public sector: linking strategic management and performance. Publ. Adm. Rev. **70**, s246–s254 (2010)
2. Johnsen, Å.: Strategic planning in turbulent times: still useful?. Publ. Policy Adm. (2022)
3. Drumaux, A., Joyce, P.: Leadership in Europe's public sector. The Palgrave Handbook of Public Administration and Management in Europe, pp. 121–139 (2018)
4. Vandersmissen, L., George, B., Voets, J.: Strategic planning and performance perceptions of managers and citizens: analysing multiple mediations. Publ. Manag. Rev. 1–25 (2022)
5. Meijer, A., Rodríguez Bolívar, M.P.: Governing the smart city: a review of the literature on smart urban governance. Int. Rev. Adm. Sci. **82**(2), 392–408 (2016)
6. Bertot, J.C., Jaeger, P.T., Grimes, J.M.: Promoting transparency and accountability through ICTs, social media, and collaborative e-government. Transform. Gov. People Process Policy **6**(1), 78–91 (2012)
7. Alshehri, M., Drew, S.: Implementation of e-government: advantages and challenges. In: International Association for Scientific Knowledge (IASK), pp. 79–86 (2010)
8. Bouton, S., et al.: How to make a city great: a review of the steps city leaders around the world take to transform their cities into great places to live and work (2013)
9. Dameri, R.P., Benevolo, C.: Governing smart cities: an empirical analysis. Soc. Sci. Comput. Rev. **34**(6), 693–707 (2016)
10. Alcaide Muñoz, L., Rodríguez Bolívar, M.P.: Different levels of smart and sustainable cities construction using e-participation tools in European and Central Asian countries. Sustainability **13**(6), 3561 (2021)

11. OECD: Strategic insights from the public governance reviews: Update. GOV/PGC (2013) 4. Public governance and territorial development directorate. Public Governance Committee. OECD, Paris (2013)
12. Meiser, J.W.: Ends+ ways+ means=(bad) strategy (2017)
13. Gil-García, J.R., Pardo, T.A.: E-government success factors: mapping practical tools to theoretical foundations. Gov. Inf. Q. **22**(2), 187–216 (2005)
14. European Energy Research Alliance: EERA joint programme on smart cities: storyline, facts and figures. J. Technol. Archit. Environ. Spec. issue **1**, 16–25 (2018)
15. Alcaide Muñoz, L., Rodríguez Bolívar, M.P., Alcaide Muñoz, C.: Political determinants in the strategic planning formulation of smart initiatives. Gov. Inf. Q. **40**(1), 101776 (2023)
16. Mintzberg, H.: The fall and rise of strategic planning. Harvard Bus. Rev. January-February (1994)
17. George, B., Walker, R.M., Monster, J.: Does strategic planning improve organizational performance? A meta-analysis. Publ. Adm. Rev. **79**(6), 810–819 (2019)
18. Bryson, J.M.: The future of public and nonprofit strategic planning in the United States. Publ. Adm. Rev. **70**, s255–s267 (2010)
19. Poister, T.H., Pitts, D.W., Hamilton Edwards, L.: Strategic management research in the public sector: a review, synthesis, and future directions. Am. Rev. Publ. Adm. **40**(5), 522–545 (2010)
20. Souki, B., Najafbeigi, R., Daneshfard, K.: The pathology of strategic planning in local organizations. J. Strateg. Manag. Stud. **12**(47), 251–275 (2021)
21. Bello-Gomez, R.A., Avellaneda, C.N.: Goal achievement in municipal strategic planning: the role of executives' background and political context. Publ. Adm. Rev. 1–20 (2023)
22. Rodríguez Bolívar, M.P., Alcaide Muñoz, C., Alcaide Muñoz, L.: Identifying strategic planning patterns of smart initiatives. an empirical research in Spanish smart cities. In: Viale Pereira, G., et al. (eds.) EGOV 2020. LNCS, vol. 12219, pp. 374–386. Springer, Cham (2020). https://doi.org/10.1007/978-3-030-57599-1_28
23. Angelidou, M.: Smart city policies: a spatial approach. Cities **41**, S3–S11 (2014)
24. Munir, R., Baird, K.: Influence of institutional pressures on performance measurement systems. J. Account. Organ. Chang. **12**(2), 106–128 (2016)
25. DiMaggio, P.J., Powell, W.W.: The iron cage revisited institutional isomorphism and collective rationality in organizational fields. In: Economics Meets Sociology in Strategic Management, vol. 17, pp. 143–166. Emerald Group Publishing Limited (2000)
26. DiMaggio, P.J., Powell, W.W.: The iron cage revisited: institutional isomorphism and collective rationality in organizational fields. Am. Sociol. Rev. 147–160 (1983)
27. Meijer, A.J., Gil-Garcia, J.R., Rodríguez Bolívar, M.P.: Smart city research: contextual conditions, governance models, and public value assessment. Soc. Sci. Comput. Rev. **34**(6), 647–656 (2016)
28. Harrison, C., Donnelly, I.A.: A theory of smart cities. In: Proceedings of the 55th Annual Meeting of the ISSS-2011, Hull, UK, September 2011
29. Collins, A., Leonard, A., Cox, A., Greco, S., Torrisi, G.: Report on urban policies for building smart cities. Project perception and evaluation of regional and cohesion policies by Europeans and identification with the values of Europe. Deliverable **4**(1) (2017)
30. Giffinger, R., Gudrun, H.: Smart cities ranking: an effective instrument for the positioning of the cities?. ACE: Archit. City Environ. UPCommons, Barcelona **4**(12), 7–26 (2010)
31. Yigitcanlar, T.: Smart city policies revisited: considerations for a truly smart and sustainable urbanism practice. World Technopolis. Rev. **7**, 97–112 (2018)
32. Esteves Araujo, J.F.F., Tejedo Romero, F.: Does gender equality affect municipal transparency: the case of Spain. Publ. Perform. Manag. Rev. **41**(1), 69–99 (2018)
33. Ward, J.H.: Hierarchical grouping to optimize an objective function. J. Amer. Statist. Ass. **58**, 236–244 (1963)

34. Ministry of foreign affairs, European Union and cooperation. https://www.exteriores.gob.es/en/PoliticaExterior/Paginas/FlujosMigratorios.aspx
35. European Commission. https://commission.europa.eu/strategy-and-policy/priorities-2019-2024/promoting-our-european-way-life/statistics-migration-europe_es
36. Millán-Franco, M., Gómez-Jacinto, L., Hombrados-Mendieta, I., García-Martín, M.A., García-Cid, A.: Influence of time of residence on the sense of community and satisfaction with life in immigrants in Spain: the moderating effects of sociodemographic characteristics. J. Community Psychol. **47**(5), 1078–1094 (2019)
37. Rodríguez-Bolívar, M.P., Alcaide-Muñoz, C., Alcaide-Muñoz, L.: Characterising smart initiatives' planning in Smart Cities: an empirical analysis in Spanish Smart Cities. In: Proceedings of the 13th International Conference on Theory and Practice of Electronic Governance, pp. 585–595 (2020)
38. Font, J., Della Porta, D., Sintomer, Y. (eds.): Participatory Democracy in Southern Europe: Causes, Characteristics and Consequences. Rowman & Littlefield, Lanham (2014)
39. Barletta, V.S., Caivano, D., Dimauro, G., Nannavecchia, A., Scalera, M.: Managing a smart city integrated model through smart program management. Appl. Sci. **10**(2), 714 (2020)
40. Hofstede, G.H., Hofstede, G.: Culture's Consequences: Comparing Values, Behaviors, Institutions and Organizations Across Nations. Sage, Thousand Oaks (2001)

Participatory Budgeting in Budapest: Navigating the Trade-Offs of Digitalisation, Resilience, and Inclusiveness Amid Crisis

Gabriella Kiss[1](✉) ⓘ, Máté Csukás[1] ⓘ, and Dániel Oross[2] ⓘ

[1] Corvinus University of Budapest, Budapest, Hungary
gabriella.kiss@uni-corvinus.hu
[2] Centre for Social Sciences, Budapest, Hungary

Abstract. In a public health crisis such as the COVID-19 pandemic, local governments had to react to external shocks. Based on previous analysis of participatory budgeting (PB) in different cities observed during the most difficult times of the pandemic in 2020 and 2021, the quality of participation and deliberation has been seriously reduced, due to the crisis. In many cases, deliberative and participatory processes were cancelled and postponed, which questioned the resilience of these processes. In other cases, online and digital solutions appeared to be a response to the health crises and their challenges. As per our previous findings, we concluded that the resilience of PB processes in the case of Budapest was increased if the resources were available to set up online platforms and create online communication. With the emerging time of crisis, such as the COVID-19 pandemic, digital government practices adopted new methods to maintain the continuation of participatory processes, while some others did not adopt new measures. These solutions were further steps towards e-participation and digitalisation of PB processes that could serve the survival of those, but they may result in less inclusive participation. To understand how resilience and inclusiveness could relate to each other, we analysed the digitalisation of PB in Budapest from that point of view. In our analysis, we explore the relationship between resilience, inclusiveness and digitalisation and the possible trade-offs between those characteristics in the case of PB in Budapest at the city level and its districts.

Keywords: e-participation · participatory budgeting · inclusiveness · resilience

1 Introduction

In a public health crisis such as the COVID-19 pandemic, local governments had to react to a series of external shocks. The quality of participation and deliberation was severely compromised by the crisis, based on previous analyses of participatory budgeting (PB) in different cities observed during the most difficult periods of the pandemic in 2020 and 2021. Municipalities are facing challenges of reassessment, accountability and learning to build future resilience in these years of uncertainty and crisis [6]. Resilience is defined

© IFIP International Federation for Information Processing 2023
Published by Springer Nature Switzerland AG 2023
N. Edelmann et al. (Eds.): ePart 2023, LNCS 14153, pp. 33–49, 2023.
https://doi.org/10.1007/978-3-031-41617-0_3

here as the ability of participatory processes and institutions to adapt and recover from disruptions, crises and changes in the social, political, economic or technological context while maintaining their core principles and objectives. The impact of the pandemic on participatory decision-making can be seen in the empirical evidence that in crisis environments caused by external shocks, governments tend to suppress opportunities for citizen participation in decision-making [5, 10]. Consultation and participatory processes have often been cancelled and postponed, calling into question their resilience. In other instances, the implementation of online and digital solutions emerged as a reaction to the health crises and the associated challenges. In our previous findings [26], we conclude that the resilience of PB processes in the case of Budapest was increased if the resources were available to set up online platforms and create online communication. With the emerging time of crisis, such as the COVID-19 pandemic, digital government practices adopted new methods to maintain the continuation of participatory processes, while some others did not adopt new measures. The implementation of these solutions marked an advancement in the realm of e-participation and digitalisation of PB processes, which were instrumental in ensuring the continuity of participatory processes during the pandemic crisis. However, it is possible that such measures have the potential to negatively impact the inclusiveness of the participatory processes. To understand how resilience and inclusiveness could relate to each other, we analysed the digitalisation of PB in Budapest from that point of view. In our analysis, we explore the relationship between resilience, inclusiveness and digitalisation and the possible trade-offs between those characteristics in the case of PB in Budapest at the city level and its districts.

2 Digital Inclusiveness and e-Participation

Digitalisation can increase inclusiveness by improving access to information and facilitating participation for people who may face barriers, such as geographical distance, mobility limitations or language barriers. New forms of participatory democracy, such as crowdsourcing, deliberation and collaborative decision-making, can also be enabled by digital platforms and tools. In the case of PB processes, e-participation can improve the effectiveness of the processes but could raise the question of digital exclusion also.

There is some evidence that e-participation in the case of PB can increase the level of participation as online access to PB dramatically reduced the costs of participation and bring more participation to decision-making [37] as citizens could participate virtually anywhere and anytime during the proposal and voting phase. Increasing the "window of time" for voting and no travelling expenses reduces the cost of participation for citizens, provoking an increase in the number of participants [37]. In that sense, the emergence of electronic ways of participation and especially deliberation can broaden the public sphere and the adoption of new technologies can increase voting turnouts and citizens' interest in politics [32]. The voting rates can be increased more by personal invitations and few voting requirements [46]. The richness and quality of information can be developed using digital solutions [29] which is inevitable for the competence of citizens in public participation [47]. It is also attractive for online solutions to be interactive and have space for discussion enabling online public debate [8, 11]. The educative function of participatory processes such as schools of democracy can reach more citizens and work as an incentive to improve their capacities and master the information necessary for their problems to make better decisions [20].

On the other hand, the expansion of the public sphere through digitalisation and online access to deliberative and participative processes requires also active participation from citizens [11]. Conversely, digitalisation can create new barriers and exacerbate existing inequalities. The design of online participation can undermine the intent to enhance inclusive participation [9]. For instance, people without access to digital technologies or poor digital skills can be excluded from online participation processes. Digital platforms may also privilege certain types of participation or certain types of participants, such as those with higher levels of social capital or technical skills. In addition, digital tools may not be able to adequately capture the diversity of perspectives and experiences within a community and may prioritise certain issues or points of view over others.

E-participation can broaden the participation divide, which implies that those, who usually participate in public issues and politics are likely to participate in participatory processes and those who are usually underrepresented in public debate remain in that position. In e-participation, the digital divide could worsen this phenomenon by being stratified by education, ethnicity, income, gender and age. In that sense, e-participation could worsen inequalities and further empower those who are already powerful and weaken those who are marginalised and even reduce their participation [21].

To prevent digital exclusion, digital tools and e-participation should be tailored to the specific needs of the target society, considering their social environment and levels of digitalisation. Various metrics, including internet penetration, broadband penetration, mobile phone use, digital literacy, eGovernment adoption, digital divide, digital infrastructure, cybersecurity, and digital entrepreneurship, are crucial for understanding digital inclusion [2, 15–18, 22–25, 35, 39, 44, 45, 48].

The case of PB in Budapest (2020–2022) underscores the need to study not only the national digital landscape but also the specific city context. As of 2022, Hungary had an 83.3% internet penetration and 74% smartphone penetration, but a below-average digital skills score of 49. In Budapest, however, digital engagement was higher, with 82.3% accessing the internet outside the home or work and 88.8% interacting with public authorities online. Yet, the digital divide remains an issue, with 35% of Hungarians aged 16–74 never having used the internet and 5% never having used a computer. Furthermore, Hungary's e-government development index, ranking 51st (0,7827) globally, might indicate challenges in online government service usage [12]. These factors must be considered when developing digital initiatives to ensure inclusivity despite apparent high digital engagement in cities like Budapest [19].

It is obvious that the mixed methods used in PB and the combination of face-to-face and online tools could increase the inclusiveness of e-participation and can create more successful processes [30, 32, 40]. There is evidence that these combinations could reduce the effects of digital exclusion [4]. But in times of crises local governments have to make critical decisions. In the time of the COVID-19 pandemic, decision-makers should decide between the need for health crisis interventions and face-to-face interactions needed for better participatory processes. In the time of social distancing, there was no chance for this kind of inclusive solutions and options. In these years digitalisation could serve as a survival for many participatory and deliberative decision-making processes. In that sense, the pandemic moves PB processes towards digitalisation to be resilient during health crises. This is especially true for PB processes in Budapest, where the introduction of the

PB coincided with the emergence of the pandemic and social distancing. Since the city is a point of reference for other local communities in the region, that is why we consider it important to examine how the PB processes in Budapest survived the pandemic, what role digitalisation played in this, and how this affected the fairness and inclusiveness of participation.

3 Participatory Budgeting in Budapest - Context and Cases

Budapest has a dual self-government system where different models of PB have been introduced in the districts and the City Council of Budapest. The cases presented in our article are in an experimental phase: PB has a maximum of 7 years of operation and no legal background. While local politicians accept theoretical arguments for promoting citizens' participation and newly elected local politicians expect to increase their party's local embeddedness by creating new contact opportunities, one of the main reasons behind introducing PB is that the process serves as a ground for experimentation [36]. From a comparative perspective, the selected cases mostly resemble the participatory modernisation model of Sintomer et al. [43] which offers consultation on public finances for citizens and gives local people a say in planning a small percentage of the total budget.

PB first appeared in Budapest in 2016, when residents in the 19th district could choose from 16 development projects determined by the municipality. In 2019, the 22nd district of Budapest also decided to provide 1 million euros to PB. After the municipal elections in 2019, the opposition parties won a majority in the General Assembly of Budapest and several districts [28], and since then the city council and several local governments (2nd, 3rd, 6th, 8th, 9th, 13th and 22nd districts) allocated a small amount (between 0.0004% and 2.5%) to PB as part of their annual budget.

Although digital tools are used everywhere, the selected cases offer a patchwork of participatory processes as different methods are used in each district and in the city council: while deliberation and face-to-face forums are integral parts of the PB process of Budapest, project submissions and voting dominate the process in the districts. To better understand how complex tools are used in the selected PB processes, we analysed 8 different cases (see Table 1).

Table 1. The analysed cases of PB process of Budapest in 2020–2022

Case	Name	Analysed years	First year of PB
Case 1	Budapest City Council	2020–22	2020
Case 2	2nd District	2021–22	2021
Case 3	3rd District	2020–22	2020
Case 4	8th District	2022	2022
Case 5	9th District	2022	2022
Case 6	13th District	2021-22	2021
Case 7	19th District	2020–22	2016
Case 8	22nd District	2020–22	2019

4 Conceptual Framework

To understand the relationship between digitalisation, resilience, and inclusiveness we built an evaluation framework based on the literature on the evaluation of e-participation processes. To analyse the 3 topics the following aspects were used.

4.1 Digitalisation

The Complexity of Digital Tools Used in PB

What is the ranking of the most used digital tools for PB according to their complexity? The more complex these platforms are, the more complex local society's IT and digital skills are required, so more citizens are excluded when more complex digital tools are used. Our assumption is that the more complex tools are used in the PB process, the fewer people can participate, therefore the sample of citizens who decide on the outcome of the budgeting process is associated with their digital skills [38].

Low-complexity tools include online surveys, social media, email, and text messaging. These tools are simple and effective for gathering participant feedback and engaging with them. Moderate complexity tools such as interactive maps, crowdsourcing platforms, and budget simulators allow for more in-depth participation and decision-making. High-complexity tools such as online deliberation platforms, augmented reality, and blockchain technology require specialized expertise and provide advanced features for constructive dialogue, visualization of projects, and ensuring transparency and security in the process [34].

The Digitalisation of Each Phase of the Process

As Sampaio [41] stated "One of the most important aspects to be examined in relation to ePB processes were related to the functions of the digital tools." The following functions were defined by Sampaio [41] 1. Engagement and mobilization, 2. Budget simulation, 3. Sending of suggestions or proposals, 4. Deliberation, 5. Voting, 6. Monitoring or assessment. Following this functionality approach, the main steps and phases of the PB processes in Budapest were identified as seen in Fig. 1.

Based on those phases the needed functions were identified and defined as criteria for evaluation. Accordingly, the following 5 criteria were defined to evaluate the digitalisation of PB processes: (1) digitalisation of communication and announcement, (2) digitalisation of proposal submission, (3) digitalisation of the voting phase, (4) Level of online deliberation and (5) digitalisation of monitoring or assessment.

4.2 Resilience

The measures of resilience were identified in our previous research [26] as responses to the crisis in 2020–21. As an evaluation criterion, we applied the characteristics that were explored during the years of Covid-19 pandemic and the time of social distancing. We explored the answers that the local government gave to this exogenous shock and investigated how the PB processes reacted to this threat. The result of our analysis of the PB processes in Budapest in 2020–2021 we identified five different types of solutions.

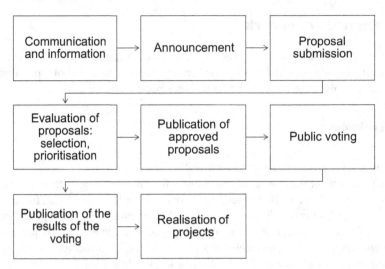

Fig. 1. General phases of participatory budgeting processes in Budapest

The five responses that were discovered in the Budapest case during the pandemic: 1. Cancellation, 2. Postponement, 3. Reduced mode, 4. Online implementation, 5. Hybrid implementation. For the evaluation of e-participation solutions, we excluded the cancellation case, as it cannot be analysed because of the fact it is not initiated at all and merged the online and hybrid implementation options. Accordingly, one criterion was identified for the evaluation of the resilience of the PB processes since they were operating in the years of the health crises by offering different solutions for the survivability of these processes.

4.3 Inclusiveness

The inclusiveness criteria are defined in the literature differently. The key issues in evaluating eParticipation platforms are the degree of transparency, interactivity and openness [1, 7, 32]. Based on the assumption that the provision of rich information and interactivity increases the transparency of decision-making. The potential for two-way communication and deliberation is also a characteristic which is emphasised [32].

The inclusiveness of the PB process is characterised by the transparency of the different phases (see Fig. 1). Based on functionality and transparency, 3 criteria were defined as the transparency of the phases: the transparency of the evaluation process, the realisation of projects and the decisions and feedback for the proposals.

Following the phases of the process, the voting phase was evaluated by its inclusiveness and fairness based on the prioritisation in voting, breakdown in topics, and the opportunity for deliberation. The criteria of the voting system were adapted.

Inclusiveness can also be defined via justice and inequality when the PB process could reflect on the participation divide and give a voice to underrepresented social groups (by age, gender, income or education). To reflect on the digital divide, solutions can be provided to handle the digital exclusion. Representing the participation and digital divide 2 criteria we used: (1) the opportunity to represent the interests of disadvantaged groups and minorities and (2) precautions against digital exclusion.

Based on the three topics (digitalisation, resilience, fairness/inclusiveness) altogether 13 criteria were developed for the framework (see Table 2).

Table 2. Evaluation framework for digitalisation, resilience and inclusiveness of PB processes

Topic	Criteria	Definition	Scale
Digitalisation	complexity of digital tools used	complex and specific tools are used or developed for the process	0 - low complexity, 1 - moderate complexity, 2 - high complexity (platform)
Digitalisation	digitalisation of communication and announcement (engagement and mobilisation)	For the announcement how developed online platforms were used for engagement and mobilisation	0-there were no online announcement and communication 1 - poor online communication (FB, website) 2 - well-established online communication
Digitalisation	digitalisation of proposal submission	for the proposal submission what level of digitalised tools were used	0 - offline proposal submission 1 - online and offline proposal submission 2 fully online proposal submission
Digitalisation	digitalisation of the voting phase	for the voting phase what kind of online and offline tools were used	0 - offline voting 1 - online and offline voting 2 fully online voting
Digitalisation	level of online deliberation	in the process what level of online deliberation or two communication channels were used (online debate, comments etc.)	0 - no online forums and space for deliberation 1 - poor options for comments 2 - online deliberation and debate was applied
Digitalisation	Digitalisation of monitoring or assessment	any option in digital tools for monitoring	0 - no online platform for monitoring of projects' realisation 1 - poor options for monitoring 2 - well developed online monitoring

(*continued*)

Table 2. (*continued*)

Topic	Criteria	Definition	Scale
Resilience	Resilience of PB process	at the time of the COVID pandemic how resilient were the PB process (cancellation is not involved in the analysis)	0 - Postponement 1 - Reduced mode 2 - Online/hybrid implementation
Fairness/inclusiveness	Transparency of the evaluation process	process transparency: the ideas are evaluated within the municipality, the process is transparent for all and how decisions are being made, they can follow the process	0-the process is not transparent, only proposals and voting lists are available, the evaluation process is invisible 1 - partly transparent 2 - fully transparent
Fairness/inclusiveness	Transparency of the realisation of projects	citizens can follow the projects' phases, are they in the realisation or planning or performing	0 - the process is not transparent 1 - partly transparent (there are a few information but not structured) 2 - fully transparent
Fairness/inclusiveness	Transparency of the decisions, feedback for the proposals	citizens can follow the proposals' phases, how the municipality decided on them and why	0 - no feedback, 1 - poor feedback (not for the individuals, no details) 2 - detailed feedback for every proposal individually

(*continued*)

Table 2. (*continued*)

Topic	Criteria	Definition	Scale
Fairness/inclusiveness	Fairness of the voting system	priorisation in voting, breakdown in topics, opportunity for deliberation	0 - single voting system (1 person one vote, no breakdown according to topics, no deliberation) 1 - partly fair voting system (priorisation in voting, breakdown in topics, no deliberation) 2 - well developed voting system (priorisation in voting, breakdown in topics, opportunity for deliberation
Fairness/inclusiveness	Opportunity to represent the interests of disadvantaged groups and minorities	the access of disadvantaged groups ensured, involvement of civil organizations, development of a topic, personal inquiry, activation, bottom-up process	0 - no detailed process to ensure the access 1 - partly ensured (topic defined to ensure equality) 2 - fully ensures (representatives are invited, topics are developed, personal invitation ensured)
Fairness/inclusiveness	Precautions against digital exclusion	any precautions against digital exclusion for example: besides online voting, there were public places where participants were able to vote or even help suggestions in writing for the ePB processes, made public voting places available, using electronic voting machines or computers with Internet access	0 - there are no precautions 1 - poorly available options 2 - there are precautions against digital exclusion

5 Methodology

Based on the literature and our previous research findings, the following research question was formulated: Is there a relationship between the scale of digital tools in the PB processes and the resilience and inclusiveness of the cases of Budapest in 2020–2022?

To answer the research question, a mixed method was used with a combination of qualitative and quantitative methods. First, qualitative research was conducted using document analysis, participant observations, and semi-structured interviews. The combination of these methods ensured the reliability of the result [33]. Data were collected systematically for every district and the Municipality of Budapest, which conducted PB in 2020–2022. During the document analysis, available information about the different processes of the districts of Budapest and the city council were collected, especially written concepts, online communication materials, newspapers, videos, and website content. Participant observation was applied in online and offline events (forums, walks, council meetings) and structured reports were created.

In the years 2020 and 2021, altogether 21 semi-structured interviews were conducted, mainly on online platforms or in person. The interviewees were selected via snowball method and purposive sampling. To gain different views about PB processes, the interviewees were representatives of local governments as organisers, experts in PB in civil society organisations, and participants in citizens' budgetary councils. The interview guide included questions about the history, different cycles (planning, deliberation, voting, implementation) and key actors of the PB process.

The evaluation of the selected PB processes followed a structured qualitative research protocol. First, written information sources were identified and collected as the transcriptions of semi-structured interviews. Based on the conceptual background, we developed a coding frame containing the evaluation criteria, their definition and how the value of each criterion should be identified (see Table 2). The researchers first prepared the data by coding the attributes of six variables related to digitalisation, six related to inclusiveness, and one related to the resilience of the process. All written material was then analysed from process to process, using the qualitative content analysis method [14, 42]. The coding was cross-checked for each process between two researchers, and scoring was discussed. The coding process involved assigning a value between 0 and 2 (0, 1 or 2) to each variable, making them ordinal variables (see Table 2). The systematic nature of the analysis allowed us to sum up the scores for each criterion within and across the analysed processes and made a comparison of the cases possible using the sum of scores [27]. After that to understand the differences between the cases, a quantitative analysis was carried out and a principal component analysis (PCA) was conducted to identify patterns and relationships among the digitalisation and inclusiveness attributes of the PB process.

6 Results

As a result of the descriptive analysis of the sum of scores in each category some conclusions can be drawn. First, from the maximum score of 16 the lowest level of scores were identified in the level of deliberation (2) and transparency (2) during the evaluation process. It can be concluded from the 8 analysed cases that the level of online deliberation of PB processes is low and PB processes are not transparent when it comes to the evaluation of the proposals. Additionally, the representation of disadvantaged groups and minorities is also at a low level in many cases in Budapest (Fig. 2).

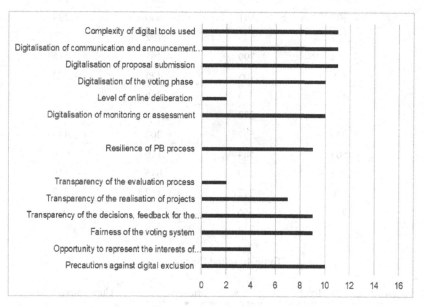

Fig. 2. Overall evaluation of the cases

Digitalisation is higher than the minimum level except for deliberation. In that sense, all the processes are digitalised in many phases of the PB. In the case of inclusiveness and fairness in Budapest cases are at a lower level except for precautions against digital exclusion, as in many cases both offline and online options were available for PB participants. The evaluation process is not transparent enough, and there is a lack of opportunities for disadvantaged groups and minorities to be represented in the processes. Altogether a higher level of digitalisation (average score 9,16) and a lower level of transparency (average score 6) can be seen from the sum of scores.

To understand the differences between the 8 cases and see a more detailed picture of the digitalisation, inclusiveness, and resilience of those, the scores were analysed in further steps. First, a principal component analysis (PCA) was conducted to identify patterns and relationships among the digitalisation and inclusiveness attributes of the PB processes. SPSS 25 software was used to carry out the PCA analysis with 12 variables (6 variables related to digitalisation, 6 related to inclusiveness), excluding the one connected to resilience. The analysis resulted in four components with eigenvalues greater than one. The first and second components were used to visualize the eight cases. The result of the analyses is detailed in Tables 3 and 4.

Component 1 is primarily characterized by the transparency of the evaluation process (0.891), inclusiveness of the voting system (0.816), and the opportunity to represent the

Table 3. Component Matrix

	Component			
	1	2	3	4
D1	,808	,399	,296	−,161
D2	−,204	,591	,640	−,293
D3	,272	,643	−,702	−,070
D4	−,244	,799	−,461	,206
D5	,719	−,105	−,425	−,372
D6	,481	,596	,286	,550
I1	,920	−,272	−,112	,212
I2	,345	,537	,208	,444
I3	,503	,584	,116	−,610
I4	,814	,119	,075	−,119
I5	,920	−,272	−,112	,212
I6	,713	−,656	,214	,004

Table 4. Initial Eigenvalues

Component	Total	% of Variance	Cumulative %
1	4,808	40,070	40,070
2	3,147	26,226	66,296
3	1,599	13,327	79,622
4	1,274	10,616	90,238

interests of disadvantaged groups and minorities (0.891). This component may represent the overall "democratic quality" of the PB processes, which emphasizes transparency, inclusiveness and inclusivity. Component 2 is primarily characterized by the digitalisation of communication and announcement (0.729), digitalisation of the voting phase (0.705), and transparency of the decisions and feedback for the proposals (0.654). This component may represent the extent to which the PB process is "digitally enabled," with an emphasis on digital communication and decision-making. In summary, the first two components of the PCA analysis were used, to analyse the association between the 8 cases in Budapest.

The factors were visualized on a scatter plot graph in SPSS (see Fig. 3), and cases were categorized, based on their resilience scores.

From Fig. 3 we could see that Budapest City Council scored high on both factors, indicating high inclusiveness and digitalisation. The 8th District scored high on inclusiveness but low on digitalisation. The 2nd District, 9th District, and 13th District scored low

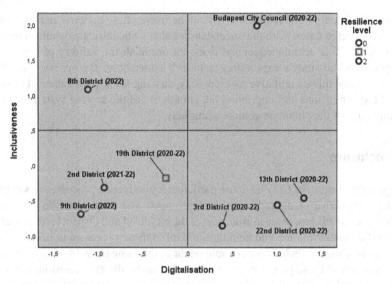

Fig. 3. Eight participatory budgeting cases of Budapest and their inclusiveness, digitalisation, and resilience scores

on both factors, indicating low inclusiveness and digitalisation. The 3rd District scored low on inclusiveness but high on digitalisation. Finally, the 19th District has average scores on both factors, indicating an average level of inclusiveness and digitalisation.

Adding the resilience score to the analysis provides further insights into how the processes performed during the COVID-19 pandemic from the perspective of their reaction and survival in the crises. Overall, the analysis shows that districts with higher inclusiveness and digitalisation scores were more resilient during the pandemic, implementing fully online/hybrid modes. In contrast, districts with lower inclusiveness and digitalisation scores were less resilient, implementing reduced modes. This could suggest that digitalisation and inclusiveness play a significant role in building resilience in public participation processes during times of crisis such as the COVID-19 pandemic. The consideration of factors such as the availability of institutional capacities and available funding can also be complementing factors since municipalities with more available resources seemed to be more resilient in those analysed cases.

7 Limitations and Validity

The study employs an experimental quantitative methodology that inherently involves trade-offs. The utilization of a 0–2 scoring system and subsequent Principal Component Analysis (PCA) is an exploratory approach to quantifying complex phenomena. This methodology, while providing a means for cross-case comparative analysis, may induce a certain degree of reductionism, potentially overlooking nuanced aspects of the phenomena under investigation. Moreover, the uneven distribution of measures across 'Digitalisation', 'Inclusiveness', and 'Resilience' dimensions further elucidates the preliminary nature of the methodology. It is essential to underline, however, that this approach was intentionally adopted to illuminate overarching patterns and correlations across multifaceted cases, with the understanding that some intricate details may be lost in the process. This acknowledgement does not diminish the validity of our findings but emphasizes the study's exploratory nature. It underscores the necessity for future research to refine this quantitative methodology, striving for a more balanced representation of the dimensions and improving the granularity of the scoring system to capture the complexity of the phenomena more accurately.

8 Conclusion

In this paper the impact of COVID-19 on participatory budgeting processes was explored from the perspective of digitalisation, inclusiveness, and resilience. We examined 8 different cases of PB processes in Budapest at the city level and district level as well and questioned if digitalisation could help the survival of those processes in times of health crises, and how digital inclusiveness could occur at the same time. In other words, how the organisation of those participatory cases could handle digital exclusion, in the time of social distancing to promote participation in a city where its initial phase coincided with the pandemic emergencies as such.

The selected three aspects, digitalisation, resilience, and inclusiveness proved to be decisive in these investigated cases as possible trade-offs were captured. In our analysis, we found a lack of resources and low administration capacity in most of the PB processes in Budapest. As for the implications of the findings, our results mirror low-level deliberation, especially in the form of online deliberation which is usual for PB processes in the East-Central European region [31]. Only city level PB process showed the existing characteristics of online and offline deliberation at all. Based on our results it can be stated that resilience requires at least a minimal level of digitalisation, cases where PB processes worked without digitisation were not resilient to the COVID-19 pandemic and thus had to be cancelled or postponed. On the other hand, high digitalisation does not go hand in hand with high inclusivity, and high inclusivity cannot be attained with a low level of digitisation. But digitalisation and its complexity are crucial characteristics also. Our study corresponds with the recent work by Annunziata [3], who found that the pandemic-induced digital shift impacted PB processes, particularly in inclusiveness and deliberation. The importance of state resources for PB adaptations and the complex effects of the digital divide were emphasized, underscoring the need for deliberate hybrid models that blend online and offline participation effectively.

As Sampaio [41] stated that the most common function of the digital tools was the "suggest and vote online" in ePB processes. We also found in the cases of Budapest that mainly voting and proposal submission are transparent and digitalised, and other aspects such as evaluation of proposals are neglected or remain in the background in a case where digital tools are complex enough to make them transparent. It is also a conclusion that next to human and digital resources the commitment of the decision-maker is also needed for a fair and inclusive process in a crisis.

Our examination of the 8 different PB processes within Budapest, varying across city and district levels, provides a rich cross-sectional analysis that illuminates a gamut of potential situations that could arise in other cities across the globe. By scrutinising the intricate dynamics within this single city, we present a valuable case study that mirrors the diverse realities of PB processes internationally. The insights gained particularly the critical role of digitalisation in enabling resilience during crisis times and its influence on inclusivity, can serve as practical recommendations for policymakers worldwide.

This study contributes to the burgeoning literature on PB and supports important discussions about effectively capitalising on digitalisation to promote inclusivity and resilience in public participation processes. Consequently, the Budapest example bears considerable relevance to an international audience, as it offers universally applicable lessons in the face of common challenges. The pandemic, while disruptive, has underlined the importance of digitalisation in public processes and has highlighted the necessity for cities worldwide to evolve and improve their participatory frameworks in the face of the ongoing digital transformation and shifting socio-political landscapes. Thus, the experiences of Budapest offer both an insightful case study and a beacon for global public participation processes seeking to optimise their digitalisation, inclusiveness, and resilience parameters.

References

1. Aichholzer, G., Allhutter, D.: Evaluation perspectives and key criteria in eParticipation. In: Proceedings of the 6th Eastern European eGovernment Days (2008)
2. Akamai. State of the Internet/Connectivity Report Q2 2020
3. Annunziata, R.: Digitalization of participatory budgeting in the context of the pandemic: the cases of San Lorenzo and Vicente López in Argentina. Local Dev. Soc. (2023). https://doi.org/10.1080/26883597.2023.2181705
4. Aström, J., Grönlund, A.: Online consultations in local government: what works, when, and why. connecting democracy: online consultation and the flow of political communication, **75** (2012)
5. Bhusal, T.: Citizen participation in times of crisis: understanding participatory budget during the COVID-19 pandemic in Nepal. ASEAN J. Community Engagement **4**(2), 321–341 (2020). https://doi.org/10.7454/ajce.v4i2.1103
6. Boin, A., Lodge, M., Luesink, M.: Learning from the COVID-19 crisis: an initial analysis of national responses. Policy Des. Pract. **3**(3), 189–204 (2020). https://doi.org/10.1080/25741292.2020.1823670
7. Cabiddu, F.: The use of web services for inclusive decision process: towards the enhancement of e-democracy. In: D'Atri, A., Saccà, D. (eds.) Information Systems: People, Organizations, Institutions, and Technologies, pp. 39–47. Springer, Cham (2009). https://doi.org/10.1007/978-3-7908-2148-2_6
8. Coleman, S., Moss, G.: Under construction: the field of online deliberation research. J. Inf. Technol. Polit. **9**(1), 1–15 (2012). https://doi.org/10.1080/19331681.2011.635957
9. Coleman, S., Sampaio, R.C.: Sustaining a democratic innovation: a study of three e-participatory budgets in Belo Horizonte. Inf. Commun. Soc. **20**(5), 754–769 (2017). https://doi.org/10.1080/1369118X.2016.1203971
10. Curato, N., Sass, J., Ercan, S.A., Niemeyer, S.: Deliberative democracy in the age of serial crisis. Int. Polit. Sci. Rev. **43**(1), 55–66 (2022). https://doi.org/10.1177/0192512120941882
11. Dahlberg, L.: The internet and democratic discourse: exploring the prospects of online deliberative forums extending the public sphere. Inf. Commun. Soc. **4**(4), 615–633 (2001). https://doi.org/10.1080/13691180110097030
12. EGDI E-Government Development Index (2023). https://publicadministration.un.org/egovkb/en-us/Data-Center

13. EITO. European IT Observatory (2020)
14. Elo, S., Kyngäs, H.: The qualitative content analysis process. J. Adv. Nurs. **62**(1), 107–115 (2008). https://doi.org/10.1111/j.1365-2648.2007.04569.x
15. EU Agency for Cybersecurity. Annual Cybersecurity Report (2020)
16. European Commission. Digital Economy and Society Index - DESI (2022)
17. European Commission. Digital Entrepreneurship Monitor (2019)
18. European Commission. Digital Economy and Society Index 2020 (2020)
19. Eurostat Regional Statistics. (2023). https://ec.europa.eu/eurostat/web/regions/data/database
20. Fung, A., Wright, E.O.: Deepening democracy: innovations in empowered participatory governance. Polit. Soc. **29**(1), 5–41 (2001). https://doi.org/10.1177/0032329201029001002
21. Goldfinch, S., Gauld, R., Herbison, P.: The participation divide? Political participation, trust in government, and e-government in Australia and New Zealand. Aust. J. Publ. Adm. **68**, 333–350 (2009). https://doi.org/10.1111/j.1467-8500.2009.00643.x
22. Helsper, E.J., Eynon, R.: Digital natives: where is the evidence? Br. Edu. Res. J. **36**(3), 503–520 (2010)
23. ITU. Global Cybersecurity Index 2020 (2020)
24. ITU. Measuring the Information Society Report (2018)
25. ITU. World Telecommunication/ICT Indicators Database (2020)
26. Kiss, G., Csukás, M., Oross, D.: Social distancing and participation: the case of participatory budgeting in Budapest, Hungary. In: Lissandrello, E., Sørensen, J., Olesen, K., Steffansen, R.N. (eds.) The New Normal in Planning, Governance and Participation, Springer, Cham (2023). https://doi.org/10.1007/978-3-031-32664-6_10
27. Kovács, E., et al.: Evaluation of participatory planning: lessons from Hungarian Natura 2000 management planning processes. J. Environ. Manag. **204**(1), 540–550 (2017). https://doi.org/10.1016/j.jenvman.2017.09.028
28. Kovarek, D., Littvay, L.: Greater than the sum of its part(ie)s: opposition comeback in the 2019 Hungarian local elections. East Eur. Polit. **38**(3), 382–399 (2022). https://doi.org/10.1080/21599165.2022.2038571
29. Loukis, E., Xenakis, A., Charalabidis, Y.: An evaluation framework for e-participation in parliaments. Int. J. Electron. Gov. **3**(1), 25–47 (2010)
30. Macintosh, A., Whyte, A.: Towards an evaluation framework for eParticipation. Transform. Gov. People Process Policy **2**(1), 16–30 (2008). https://doi.org/10.1108/17506160810862928
31. Marczewska-Rytko, M., et al. (eds.): Civic Participation in the Visegrad Group Countries after 1989. Maria Curie-Skłodowska Press, Lublin (2018). ISBN 978-83-227-9101-1
32. Medaglia, R.: eParticipation research: moving characterization forward (2006–2011). Gov. Inf. Q. **29**(3), 346–360 (2012). https://doi.org/10.1016/j.giq.2012.02.010
33. Miles, M.B., Huberman, M.A., Saldana, J.: Qualitative Data Analysis: A Methods Sourcebook SAGE Publications, Thousand Oaks (2018)
34. Mkude, C., Pérez-Espés, C., Wimmer, M.A.: Participatory budgeting: a framework to analyze the value-add of citizen participation. In: Proceedings of the Annual Hawaii International Conference on System Sciences, pp. 2054–2062 (2014). https://doi.org/10.1109/HICSS.2014.260
35. OECD. OECD Broadband Statistics (2020)
36. Oross, D., Kiss, G.: More than just an experiment? Politicians arguments behind introducing participatory budgeting in Budapest. Acta Polit. **58**, 552–572 (2023). https://doi.org/10.1057/s41269-021-00223-6
37. Peixoto, T.: Beyond theory: e-Participatory budgeting and its promises for eParticipation. Eur. J. ePract. **7**(5), 1–9 (2009)

38. Peixoto,T., Steinberg, T.: Citizen Engagement Emerging Digital Technologies Create New Risks and Value, World Bank Group.pdf (2018). https://documents1.worldbank.org/curated/en/907721570027981778/pdf/Citizen-Engagement-Emerging-Digital-Technologies-Create-New-Risks-and-Value
39. Pew Research Center. Mobile Technology and Home Broadband 2021 (2021)
40. Sæbø, Ø., Rose, J., Molka-Danielsen, J.: eParticipation: designing and managing political discussion forums. Soc. Sci. Comput. Rev. **28**(4), 403–426 (2010). https://doi.org/10.1177/0894439309341626
41. Sampaio, R.C.: e-Participatory budgeting as an initiative of e-requests: prospecting for leading cases and reflections on e-Participation. Revista De Administração Pública **50**(6), 937–958 (2016). https://doi.org/10.1590/0034-7612152210
42. Schreier, M.: Qualitative content analysis. In: Flick, U. (ed.) The SAGE Handbook of Qualitative Data Analysis, pp. 170–183. Sage Publications, London (2014). https://doi.org/10.4135/9781446282243
43. Sintomer, Y., Röcke, A., Herzberg, C.: Participatory Budgeting in Europe: Democracy and Public Governance. Routledge, London (2016). https://doi.org/10.4324/9781315599472
44. Statista. Internet, broadband and mobile internet penetration in Hungary from January 2021 to January 2022 (2022)
45. United Nations. E-Government Survey 2020 (2020)
46. van der Does, R., Bos, D.: What can make online government platforms inclusive and deliberative? A reflection on online participatory budgeting in Duinoord, The Hague. J. Deliberative Democracy **17**(1), 48–55 (2021). https://doi.org/10.16997/jdd.965
47. Webler, T.: Right discourse in citizen participation: an evaluative jardstick. In: Renn, O., Webler, T., Wiedermann, P. (eds.) Fairness and Competence in Citizen Participation: Evaluating Models for Environmental Discourse. Kluwer Academic, Dordrecht (1995)
48. World Bank. World Development Report 2019

Residents' Voices on Proposals

Analysing a Participatory Budgeting Project in Seoul Using Topic Modelling

Bokyong Shin[(✉)] [iD]

University of Helsinki, Helsinki, Finland
`bok-yong.shin@helsinki.fi`

Abstract. In participatory budgeting, citizens submit budget proposals for funding to improve their neighbourhoods. These proposals are publicly accessible and are a crucial source for identifying local problems and preferences. However, they are challenging to comprehend due to the extensive amounts of data involved. This article fills this gap by applying the structural topic model as an automated content analysis method for identifying the major topics and trends in the case of participatory budgeting in Seoul, South Korea. In total, 26,131 proposals submitted from 2013 to 2022 were analysed. The result shows that citizens were concerned about 12 topics under three themes: facility maintenance, community rebuilding, and risk prevention. While community rebuilding topics are decreasing, residents have paid increasing attention to public facilities, traffic, and pollution problems, reflecting topical changes over time. This study contributes by demonstrating the applications of an automated content analysis of extensive citizens' inputs in democratic processes.

Keywords: Participatory budgeting · proposal · Topic modelling · Structural topic model · Seoul · Open data

1 Introduction

Public authorities increasingly enhance opportunities for citizens to participate, deliberate, and influence policy-making processes [1]. For instance, participatory budgeting (PB), a process in which citizens engage in public budgeting [2], was implemented in over 4,000 cases globally in 2020 despite the pandemic [3]. In the digital era, many of these practices occur in online platforms, arousing interest in utilising citizen-generated data to assess democratic processes [4–7].

In this article, citizens' budget proposals are of central interest. Citizens may directly initiate proposals for improving neighbourhoods to be decided by popular vote in PB. Since budget proposals are crucial for understanding citizens' voices to be reflected in deliberation and decision-making, previous studies have conducted content analyses, albeit by using manual coding schemes that need substantial human resources to read

© IFIP International Federation for Information Processing 2023
Published by Springer Nature Switzerland AG 2023
N. Edelmann et al. (Eds.): ePart 2023, LNCS 14153, pp. 50–64, 2023.
https://doi.org/10.1007/978-3-031-41617-0_4

and annotate data [8–10]. The recent advancements in natural language processing can help overcome the limitation by automatically detecting common themes and trends, but empirical studies are still scarce [11, 12].

This article fills the gap by employing the structural topic model (STM) [13, 14], a type of unsupervised learning model as an automated content analysis method for analysing budget proposal data. Contrary to the supervised learning model, topic modelling does not require prior labelling of proposals [12], which suits the current study that examines under-explored proposal data in PB. This article raises the following two research questions: (1) What have the main topics of budget proposals been over the last decade? (2) How has the prevalence of topics changed over time?

This article demonstrates the application by focusing on the case of PB in Seoul, South Korea (hereafter Korea). Korea is one of the few countries that have mandated PB at the municipal level since 2011, and the Seoul PB has been the most active case with ample proposal data. This article examined 26,131 proposals submitted from 2013 to 2022 with STM, to understand what residents proposed over the last decade. Based on the results, this article concludes by discussing the suitability of topic modelling for PB research.

2 Participatory Budgeting and Citizen Proposals

PB is a process of involving non-elected citizens in the spending of public budgets [2]. Contrary to the traditional top-down budgeting in representative democratic systems, PB is characterised by a combination of top-down and bottom-up approaches as citizens can directly engage in initiating, deliberating, deciding, implementing, and monitoring, or all budgeting processes [2, 15]. Originating from Porto Alegre in Brazil in the late 1980s, PB has been diffused to Africa, North America, Europe, and Asia [3]. The global diffusion has attracted diverse academic interests [16]: from institutionalisation and antecedents of PB to its processes, designs and outcomes.

Previous studies have recently considered citizen proposals as empirical sources for identifying prominent themes and budget allocation patterns [8–10, 17, 18]. On the one hand, local governments collect citizen proposals and make them available to the public for deliberation and voting, increasing the accessibility of proposal data. However, on the other hand, the increase in the scale of citizen inputs poses "information overload" problems for understanding and synthesising multiple voices [19]. As Fishkin [20] noted, mass participation is a cornerstone of democracy, but it requires a fair and transparent system for mass content. Therefore, a fundamental step for PB is to identify what citizens want and propose for improving their neighbourhoods.

Previous studies addressed the problem based on pre-defined frameworks. For instance, Falanga et al. [8] followed nine themes categorised by the public authority for studying PB proposals under the theme "environment, green structure, and energy" in the case of PB in Lisbon. This approach is useful in a practical sense as many cases let citizens or experts classify submitted proposals. However, a potential problem is the low level of accuracy and the lack of control for research. Other studies deductively defined themes *priori* and employed human coders who read, understood, and labelled proposals [9, 18]. The manual coding scheme is a well-established method in content analysis with standardised quality checks (e.g., inter-coder reliability), but it lacks the ability to detect emerging topics and handling extensive data [21]. This article suggests topic modelling as an inductive methodological approach, discussed in the method section.

3 Participatory Budgeting Project in Seoul

Korea is one of the few countries to have mandated PB by national law [3]. Accordingly, 243 municipalities (eight metropolitan cities, nine regional governance, 75 cities, 82 counties, and 69 districts) have implemented PB, among which this article focuses on the case of Seoul.

Seoul is the capital city of Korea. The size of the land is only 0.6% of the national territory [22] but accounts for 22.7% of the total GRDP (gross regional domestic product) [23] and houses 9.5 million people, 18.1% of the total population [24]. Residents elect a mayor and council members every four years, constituting a mayor-council form of city government [25]. Compared to the council-manager form (appointed mayor), an elected mayor has strong leadership and policy initiatives to prepare an annual budget [26]. The amount of the annual budget for the Seoul Metropolitan Government in 2021 was 58 trillion KRW (42 billion Euros), 16.2% of the total municipal budget, allocated to social welfare (41%), administration (14%), transportation (10%), education (7%), and the environment (5%) [27]. The budgeting process in Seoul follows three stages in a year-long iterative cycle, as shown in Fig. 1 [28]:

- Budget planning: The City establishes a mid-term financial plan and reviews investments and loans to create a budgeting guideline.
- Budget compilation: Based on the guideline, each city department (or task unit) submits budget requests to the Planning and Cooperation Office, which compiles them. The city then prepares a budget internally while negotiating with the city council and national government. The mayor submits the budget to the city council for approval in November.
- Budget resolution: The city council deliberates and resolves the budget in December.

In 2012, the Seoul government established the Seoul PB, called the Citizen Participatory Budgeting System (CPBS), by enacting the Seoul Metropolitan City Participatory Budgeting Management Ordinance under the leadership of left-wing Mayor Won-Soon Park [29]. The basic idea of CPBS is to allow citizens to engage directly in the budgeting process to enhance budgetary transparency and democracy [30, Article 3]. Since there was little room for public participation in the traditional budgeting process, public officials and citizens have learnt through trial and error, reflected in 13 revisions of the ordinance over the last decade.

Despite the frequent changes, the two main programs under CPBS are the Proposal and the Deliberation programs [31]. The Proposal program (*je-an-hyeong* in Korean) encourages residents to submit proposals at different levels (e.g., city, borough, and district). City experts then screen the proposals to check their feasibility and legality in April. Citizens choose the filtered proposals via a popular vote held in August. Likewise, the program's main feature is that citizens play initiators and decision-makers in a pre-defined budget ceiling. Similar cases can be found in New York [32], Helsinki [4], and Polish cities [33].

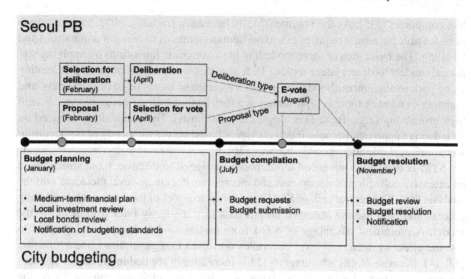

Fig. 1. Citizen Participatory Budgeting System

While the Proposal program allows citizens to voice their concerns regarding matters in everyday life, deliberation is limited. The Deliberation program (*sug-ui-hyeong*) was established in 2019 to encourage citizens to engage in more focused topics and existing city budgeting processes from a long-term perspective [31]. Citizens, experts, and public officials join the citizen participatory budgeting committee, which consults the city's mid and long-term budgeting and investment plans. Moreover, the committee reviews and selects citizen-generated proposals (city-wide level) for a popular vote in April.

The mayor encloses the results of CPBS when submitting a budget bill to the city council according to the *Local Finance Act*, Article 39 (revision, 2018).

4   Data and Methods

This article has an analysis of all budget proposals submitted between 2013 and 2022, to understand the main themes and trends over the last decade. This article has not included the Deliberation program because it operated only for a limited period (2019–2021). The proposal data are open and accessible via an application programming interface (https://opengov.seoul.go.kr/data/19195057). In addition, a custom scraper collected the data from 2020 to 2022 on the website (https://yesan.seoul.go.kr/) to obtain supplementary information. The final dataset contained 26,131 budget proposals with proposal-level variables, including ID, year, duration, location, budget, and category. Private information was not collected. KoNLPy, a Python library for Korean natural language processing, was used to delete numbers and special characters and collect words (nouns and verbs) using a Part-of-Speech tagger (*Okt* tagger).

The structural topic model (STM) [13, 34] was employed to analyse the main topics of proposals. Content analysis has traditionally relied on human coders who read, understand, and interpret texts, but "coders are humans even when they are asked to act like computers" [35]. As the volume of texts increases, manual coding schemes cease to be feasible because it requires extensive human resources to conduct annotations and labelling. The basic idea of topic modelling is to overcome limitations by applying statistical models to detect latent topics [3]. It assumes that each proposal (e.g., Creating a dog park in a neighbourhood) consists of a mixture of topics (e.g., ecology, pets, and planning), and each topic is represented as a distribution over co-occurring words with high probability (e.g., Topic *Pets*: dogs, cats, and birds). Topic modelling is based on an inductive approach because it detects topics based on the frequency of co-occurring words in a scalable and reproducible way [36].

STM is one of the advanced topic models designed to discover topic structures in an extensive collection of documents and incorporate document-level metadata into the analysis [14]. The central task of STM is to define a model of the document-generating process and then use the observed corpus data to statistically infer parameters in the model. A prominent advantage of STM from earlier models is that it allows one to test the effect of document-level covariates on topics in a generalised linear function [13, 14]. Because of this advantage, STM is increasingly applied in the social sciences for hypothesis testing using various types of textual data, including interviews, online forums, social media, and academic literature [21]. In this article, the STM was employed to examine whether topic prevalence has changed significantly over time. Model outputs are in Korean, translated into English by the author to report findings in this article.

5 Findings

5.1 Descriptive Statistics

Figure 2 shows the main themes of the proposals. The city government defined categories (categories have been changed several times) and then asked citizens to choose the most appropriate theme when submitting their proposals. The result shows that citizens are primarily concerned about the environment (24.1%) in their proposals. Other themes were welfare and health (16%), culture, sports, tourism (12.4%), and transportation (11.3%). While these categories are helpful for overviewing proposal data, it is still challenging to understand what topics citizens primarily proposed. For instance, the environment category is too broad, involving diverse activities for natural ecosystems, living, and built environments. Another problem is that it depends on subjective judgments by citizens. If a proposal is about organising community activities for cleaning graffiti off the wall, for instance, the proposal could be classified as an environment, culture, or housing theme, depending on different viewpoints.

Citizens also filled in targeted areas of each proposal (multiple areas are allowed), which Fig. 3 shows popular boroughs (*gu* in Korean) on the map. The greener a borough indicates that more proposals targeted the area. The GRDP per capita was also marked to show the differences in the size of each borough's economy. A larger circle indicates a more vibrant regional economy. The result of a Pearson correlation coefficient between 'the number of proposals per capita' and 'GRDP per capita' yielded that there was little evidence that proposals targeted (dis)advantaged areas ($r(23) = -.3, p > .15$). Rather, Fig. 3 implies that citizens were more interested in central areas when they initially submitted proposals. One reason is that the Seoul PB promoted equality (redistribution for

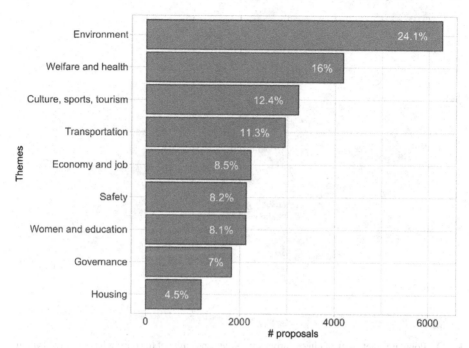

Fig. 2. Themes of budget proposals

everyone) rather than equity (redistribution for disadvantaged groups) in participation, unlike other countries [37]. However, No and Hsueh [37] found that the budget was finally allocated more to disadvantaged districts by subcommittees.

The number of proposals submitted is a simple and helpful barometer of public engagement. In 2013, residents submitted 1,460 proposals, which substantially increased for the next three years to a peak of 3,824 proposals in 2016. However, since then, it has continuously decreased, reaching the lowest number of 474 proposals in 2022. While a red line in Fig. 4 Shows this decreasing participation trend, residents increasingly asked for more funding per proposals. Ironically, 2022 shows the lowest number of proposals with the highest requested budget per proposal on average, 734 million KRW (around 0.5 million Euros).

It is noticeable in Fig. 4 that while the budget ceiling is stable, shown as a blue line, the amount of approved budget by the city council has declined since 2017. In 2022, the city council passed only five proposals with 4.5% of the budget ceiling [38]. Another crucial point is a significant decline in all figures in 2022. While more years would be required to trace this new trend, a critical event in 2021 was that Se-Hoon Oh from the right-wing party took over as the mayor after a lengthy tenure by a left-wing mayor (2011–2020). In 2022, mayor Oh ceased all borough (*gu*) and district (*dong*) level programs while focusing on three categories (transportation, health, and environment) at the city-wide level.

Overall, 2022 was the tenth anniversary of Seoul PB. As No [39] was previously concerned about its future under a new political leadership, this result reveals that the

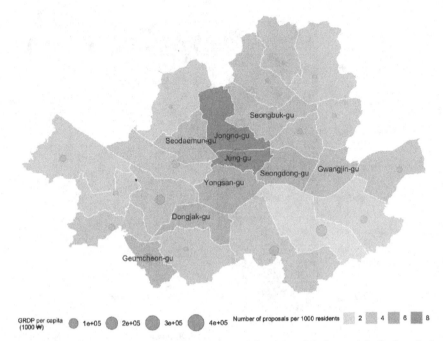

Fig. 3. Popular target areas of citizen proposals Note: The name of the borough is displayed when the value of the number of proposals per 1000 residents is higher than the average.

swing of a political pendulum between left and right can still influence institutionalised PB, especially under a mayor-council form of city government. A further in-depth investigation is required to examine critical contextual factors, such as deteriorating citizen participation and lack of communication, in addition to political leadership.

5.2 Topic Modelling

The Number of Topics. This article used two diagnostic measures to determine the optimal number of topic models—exclusivity and semantic cohesiveness [14]. First, exclusivity is higher when the most frequent words in a topic are unlikely to appear in other topics, so that each topic is unique. Second, semantic cohesiveness is higher when most frequent words in a topic also appear in other topics, making topics semantically consistent. These metrics are typically in a trade-off, so choosing topic numbers with high values for both [14]. Based on manual inspection and the metrics, the 12-topic model was finally chosen for the analysis.

Topic Interpretation. Table 1 shows the results of the 12-topic model. The third column (Keywords) presents ten keywords of each topic determined by the most frequent words within a topic, and the fourth column (%) shows the expected topic proportion in the entire corpus. While the second and third columns are direct model outputs, the second column (Topics) presents a summary of topics based on manual inspections by the author. Likewise, topic modelling also requires substantial domain knowledge for coding (topic labelling) and qualitative interpretation (description of topics) [36].

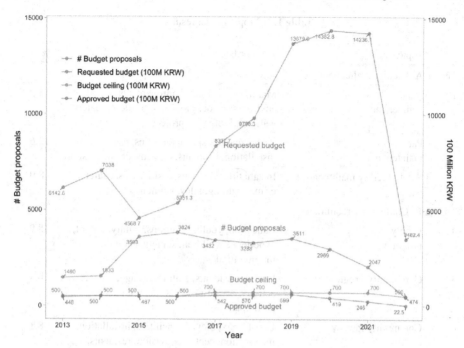

Fig. 4. Annual budget proposal numbers and requested, ceiling, and approved budgets

An investigation of model outputs and representative proposals revealed that citizens in Seoul were primarily concerned with facility maintenance, community rebuilding, and risk prevention. The first prominent theme was about maintaining damaged roads and footpaths (Topic 1), park and sports facilities (Topic 2), and safety facilities (Topic 3). It implies that many residents conceived PB as a channel for asking for public services or seeking help to address the daily problems they experienced.

The second theme was community rebuilding. According to the Survey on Urban Policy Index [40] in 2021, seven out of ten citizens in Seoul responded that "there is a person who can get help in times of need". However, trust in family, friends, public organisations, neighbours, strangers, and foreigners (in descending order) has declined over the last five years. Considering the contextual background, the topic model result indicates that citizens want to spend public resources to restore their community and neighbourhoods. Topics 4 (Family programs), 5 (Community programs), and 6 (Community facilities) were about facilitating educational and cultural programs for restoring families and communities. Topics 7 (Youth employment) and 8 (Welfare programs) focused on disadvantaged social groups, including the youth, the elderly, and the disabled. In Topic 9 (Boosting tourism), residents proposed ideas for boosting tourism to revitalise neighbourhoods.

The third theme was risk prevention. According to the Survey on Urban Policy Index [40], in 2021, citizens in Seoul sensed risk mostly from the pandemic, unemployment, social conflict, corruption, and cyberbullying (in descending order), while concerns about traffic and safety were below the average. However, citizens increasingly felt more at

Table 1. 12 topic model results

#	Topics	Keywords	%
Theme A: Facility maintenance			
1	Road maintenance	Maintenance, road, improvements, environment, deterioration, residents, safety, walking, footpaths, project	13.3
2	Park maintenance	Park, creation, space, facilities, use, installation, residents, citizens, children, sports	12.6
3	Safety facility maintenance	Installation, area, bus, resident, use, safety, crime, lighting, alley, women	8.9
Theme B: Community rebuilding			
4	Family programs	Education, youth, programs, family, school, experience, children, society, parents, multicultural	8.9
5	Community programs	Region, residents, culture, village, space, activity, operation, art, community, participation	8.4
6	Community facility	Use, library, bicycle, facilities, installation, information, centre, provision, residents, operations	8.1
7	Youth employment	Education, society, youth, business, job, support, operation, employment, region, start-up	7.3
8	Welfare programs	The elderly, the disabled, society, welfare, health, support, project, service, provision, household	7.1
9	Boosting tourism	Seoul, market, tourism, street, history, culture, region, tradition, revitalisation, tourists	6.8
Theme C: Risk prevention			
10	Traffic safety measures	Safety, installation, vehicle, accident, child, traffic, road, crosswalk, parking, pedestrians	6.6
11	Pollution prevention	Fine dust, environment, garbage, installation, energy, business, citizens, recycling, eco-friendly, collections	6.4
12	Accident prevention	Management, prevention, safety, outbreak, project, situation, citizen, problem, implementation, housing	5.6

risk from traffic and accidents than from other forms of risk. While pollution was not among the risks outlined in the survey, citizens responded that they are increasingly concerned about environmental activities, including green space, fine dust, regulation, and the green economy. The model result reflects the recent trend in traffic safety (Topic 10), pollution prevention (Topic 11), and accident prevention (Topic 12).

Longitudinal Change of Topic Prevalence. As society changes, the topicality of proposals might also change. I investigated the temporal dynamics of topic prevalence over the last decade according to the three themes identified, as shown in Figs. 5, 6, and 7. The graphs plot a smooth function of year (median) with 95% confidence intervals in dotted lines [34]. Before the investigation, it is worth recalling that under the new mayor, the Seoul government started to retract the program in 2022. Therefore, citizens could only submit proposals on three thematic areas: transportation, health, and the environment. As a result, most topics significantly declined in 2022, except those in the thematic areas.

First, Fig. 5 shows the longitudinal change in topic proportion by year in Theme 1 (Facility maintenance) with mixed results. While Topic 1 (Road maintenance) is the most prominent topic within the corpus (as shown in Table 1), it has a decreasing trend. The rationale for this trend requires further investigation: maintenance issues of roads and footpaths might have been resolved over time, or citizens might have given up asking in proposals. In contrast, residents increasingly shift their attention to parks (e.g., the creation of the park, space for children, and sports facilities) and safety facilities (e.g., for buses, crime, lighting, alley, and women). In 2022, the three topics significantly declined due to the government's intervention.

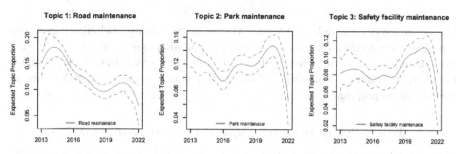

Fig. 5. Longitudinal expected topic proportions (Theme 1: Facility maintenance)

Second, topics in Theme 2 (Community rebuilding) show a decreasing trend until 2021, as shown in Fig. 6. Topics 4 (Family programs), 5 (Community programs), 7 (Youth employment), 8 (Welfare programs), and 9 (Boosting tourism) were all declined from 2016. While this result calls for a further in-depth investigation, one reason could be that rebuilding a community is more complex than fixing the roads because the sense of community resides in social bonds, trust, and support—the intangible asset of social capital. As shown in Table 1, citizens proposed educational and cultural programs as solutions, but they require long-term efforts for long-term outcomes. Figure 6 shows that these topics had initially increased until 2016, then continuously decreased except for Topic 6 (Community facility), which concerns the use of community facilities (e.g., library, bicycle, and cultural centres). It is also worth noting that Topic 8 (Welfare programs) increased substantially in 2022 due to the thematic focus.

Third, topics in Theme 3 (Risk prevention) have continuously increased, as shown in Fig. 7. Topic 11 (Pollution prevention) has increased from almost zero topic proportion in

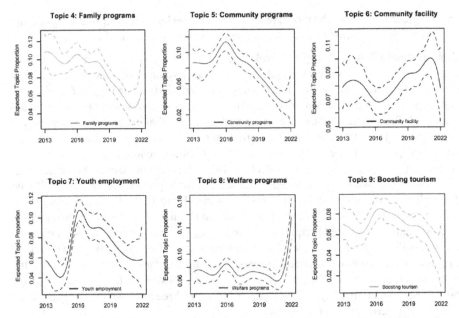

Fig. 6. Longitudinal expected topic proportions (Theme 2: Community rebuilding)

2013 to the top in 2022, reflecting residents' growing concern over pollution prevention. It is also worth noting a mixed trend of two transportation topics, a decreasing Topic 1 (Road maintenance) while increasing Topic 10 (Traffic safety measures). Table 2 displays the titles of example proposals highly associated with the topics to highlight the different contents of the proposals.

Table 2. Example proposal titles highly associated with Topics 1 and 10.

Topic 1 (Road maintenance)	Topic 10 (Traffic safety measures)
"The stones of Mountain A pour down." "Please repair the old sewer pipes and the broken road." "Please fix the old, badly damaged pavement."	"Crossroads tick-tock. There is no more tailgating!!!" "Installation of crossing floor signals for pedestrian safety in front of elementary schools." "CCTV installation to crack down on illegal parking, such as in child protection areas."

6 Discussion

This article examined the contents of residents' budget proposals (2013–2022) in a case of participatory budgeting in Seoul. The Seoul PB was selected for the current study because it is one of the most active cases in Korea, with 26,131 citizen-generated

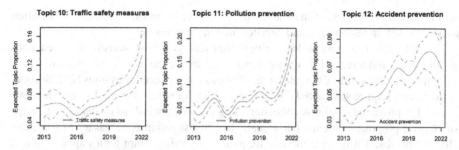

Fig. 7. Longitudinal expected topic proportions (Theme 3: Risk prevention)

proposals. The structural topic model [13, 14] was employed to identify the main topics proposed and their longitudinal trends. This section discusses the findings by answering the two research questions.

The first question was about the major topics of budget proposals. The results revealed 12 major topics under three themes. The first theme consists of three topics that concern the maintenance of roads (Topic 1), parks (2), and safety facilities (3). These topics are mundane but are probably the most frequent daily problems citizens experience. The second theme consists of topics on community rebuilding, such as family (Topic 4), community (5, 6), youth (7), welfare (8), and tourism (9) through educational and cultural programs. The third theme was risk prevention measures, including traffic safety (Topic 10), pollution (11), and accidents (12). In sum, three major themes in budget proposals were facility maintenance, community rebuilding, and risk prevention.

The second question was about the longitudinal change of topic prevalence. While most topics in the facility maintenance and the risk prevention themes showed an upward trend, topics in the community rebuilding theme have become less popular over the last decade, as shown in Figs. 5, 6 and 7. The mixed trend calls for an in-depth investigation of the contextual background. Nevertheless, given the year-long budgeting cycle, one explanation is a growing tendency to favour proposals for short-term processes and outputs (e.g., facility maintenance and risk prevention) over long-term processes and outcomes (e.g., community rebuilding).

7   Conclusions

7.1   Advantages and Disadvantages of Topic Modelling

Despite the potential of natural language processing techniques in analysing the content of extensive citizens' inputs, empirical research is still rare [11]. This article contributes by demonstrating the applications of topic modelling (structural topic model) in analysing citizen proposals for participatory budgeting. An advantage of this approach is that it systematically reveals topic structures, trends, and hypothesis testing, which is essential for social science research [21]. Topic modelling could be applied to different PB (or participatory) cases in single or comparative research designs to understand what citizens propose in different contexts. Contrary to manual coding, topic modelling produces timely and reproducible information for practitioners to identify citizens' voices *during* the PB process based on better understandings of themes, trends, and polarisation to be reflected in deliberation and decision-making processes.

Nevertheless, topic models have critical disadvantages. First, statistically based models like topic models tend to ignore word order in sentences (c.f. the Bag-of-Words assumption), which limits the interpretability of more nuanced meanings [21]. Second, topic models require substantial domain knowledge for researchers and readers to interpret and improve the results. A collection of co-occurring words as model outputs helps summarise proposal contents, but a qualitative investigation is necessary to contextualise the results. Third, topic models are poor at detecting minor topics appearing in a few documents. It is a significant limitation because PB, in principle, aims to reflect the voices of marginal groups or disadvantaged areas [2, 15].

7.2 Limitations and Future Implication

There are limitations to the findings. First, this article analysed citizen proposals submitted to the Seoul PB; thus, the result is not generalisable to other cases in different cities and countries. Second, 2021 was an important transition year of political leadership in Seoul, resulting in a dramatic decline of the program, as shown in Fig. 4. While it suggests dropping the data of 2022 to improve model results, it was included to show how PB originated from leftism remains as a political strategy rather than "a politically neutral device" [41] even in a legally mandated context. Like the original Brazilian model, PB in Korea was initiated by a minor left-wing party and civil society organisations until the left-wing national government picked up the idea. Seoul's case shows that a top-down and mandatory PB initiative does not automatically guarantee sustainability without bottom-up initiatives and supports. Other countries and cities under the process of institutionalisation could take notes from the Seoul case. Third, this article used fundamental functions of topic modelling, while not fully demonstrating many useful applications.

Future research would extend the current study by 1) increasing topic numbers to investigate more detailed topic structures, 2) incorporating other proposal-level covariates (e.g., proposal duration, area) to test their effect on topics, 3) measuring topic correlations to identify central themes and topic clusters, 4) comparing topics submitted by citizens and model results, or 5) combining with voting and budget data to examine the allocation of budgets.

Topic models for detecting marginal voices or conflictual issues also provide promising research avenues. In this case, future research could utilise topic modelling or qualitative data (e.g., interviews) to obtain local knowledge and sequentially feed it into semi-supervised topic models by weighing anchor words to detect topics of interest [12]. The current study will provide a reference for future research.

References

1. Elstub, S., Escobar, O. (eds.): Handbook of Democratic Innovation and Governance. Edward Elgar Publishing, Cheltenham, UK and Northampton, USA (2019)
2. Sintomer, Y., Herzberg, C., Röcke, A., Allegretti, G.: Transnational models of citizen participation: the case of participatory budgeting. J. Publ. Deliberation. **8**, 1–9 (2012)

3. Dias, N., Enríquez, S., Cardita, R., Júlio, S.: Participatory Budgeting World Atlas 2020–2021. Oficina, Portugal (2021)
4. Shin, B., Rask, M., Tuominen, P.: Learning through online participation: a longitudinal analysis of participatory budgeting using Big Data indicators. Inf. Polity. **27**, 517–538 (2022)
5. Edelmann, N., Höchtl, B.: e-Participation in Austria: digital agenda Vienna. In: Randma-Liiv, T. Lember, V. (eds.) Engaging Citizens in Policy Making, pp. 225–243. Edward Elgar Publishing (2022)
6. Höchtl, J., Parycek, P., Schöllhammer, R.: Big Data in the policy cycle: policy decision making in the digital era. J. Organ. Comput. Electron. Commer. **26**, 147–169 (2016)
7. Johannessen, M.R., Berntzen, L.: A decade of eParticipation research: an overview of the ePart conference 2009–2018. In: Panagiotopoulos, P., et al. (eds.) ePart 2019. LNCS, vol. 11686, pp. 3–14. Springer, Cham (2019). https://doi.org/10.1007/978-3-030-27397-2_1
8. Falanga, R., Verheij, J., Bina, O.: Green(er) cities and their citizens: insights from the participatory budget of Lisbon. Sustainability **13**, 8243 (2021)
9. Szczepańska, A., Zagroba, M., Pietrzyk, K.: Participatory budgeting as a method for improving public spaces in major polish cities. Soc. Indic Res. **162**, 231–252 (2022)
10. Kołat, K., Furmankiewicz, M., Kalisiak-Mędelska, M.: What are the needs of city Dwellers in terms of the development of public spaces? A case study of participatory budgeting in Częstochowa, Poland. Int. J. Environ. Res. Publ. Health **19**, 5171 (2022)
11. Beauchamp, N.: Modeling and measuring deliberation online. In: Foucault, B., González-Bailón, S. (eds.) The Oxford Handbook of Networked Communication, pp. 322–349. Oxford University Press, New York (2018)
12. Romberg, J., Escher, T.: Automated topic categorisationof citizens' contributions: reducing manual labelling efforts through active learning. In: Janssen, M., et al. (eds.) EGOV 2022. LNCS, vol. 13391, pp. 369–385. Springer, Cham (2022). https://doi.org/10.1007/978-3-031-15086-9_24
13. Roberts, M.E., Stewart, B.M., Tingley, D., Airoldi, E.M.: The structural topic model and applied social science. In: Advances in Neural Information Processing Systems Workshop on Topic Models: Computation, Application, and Evaluation, pp. 1–20. Harrahs and Harveys, Lake Tahoe (2013)
14. Roberts, M.E., et al.: Structural topic models for open-ended survey responses. Am. J. Polit. Sci. **58**, 1064–1082 (2014)
15. Baiocchi, G., Ganuza, E.: Participatory budgeting as if emancipation mattered. Polit. Soc. **42**, 29–50 (2014)
16. Bartocci, L., Grossi, G., Mauro, S.G., Ebdon, C.: The journey of participatory budgeting: a systematic literature review and future research directions. Int. Rev. Adm. Sci. 00208523221078938 (2022)
17. Schneider, S.H., Busse, S.: Participatory budgeting in Germany–a review of empirical findings. Int. J. Publ. Adm. **42**, 259–273 (2019)
18. Rask, M., Ertiö, T.-P., Tuominen, P., Ahonen, V.: The Final Evaluation of the City of Helsinki Participatory Budgeting: OmaStadi 2018–2020. Ministry of Justice and BIBU Project Publications, Helsinki (2021)
19. Arana-Catania, M., et al.: Citizen participation and machine learning for a better democracy. Digit. Gov. Res. Pract. **2**, 1–22 (2021)
20. Fishkin, J.S.: When the People Speak: Deliberative Democracy and Public Consultation. Oxford University Press, Oxford (2009)
21. Wesslen, R.: Computer-assisted text analysis for social science: topic models and beyond. arXiv preprint arXiv:1803.11045 (2018)
22. Statistics Korea: National land status (by administrative district, by owner, by category). https://www.index.go.kr/unity/potal/main/EachDtlPageDetail.do?idx_cd=2728

23. Statistics Korea: Gross Regional Domestic Product. https://index.go.kr/potal/main/EachDt lPageDetail.do?idx_cd=1008. last Accessed 09 July 2022
24. Statistics Korea: Key population indicators. https://www.index.go.kr
25. Kim, Y.: Do council-manager and mayor-council types of city governments manage information systems differently? An empirical test. Int. J. Publ. Adm. **26**, 119–134 (2003)
26. Coate, S., Knight, B.: Government form and public spending: theory and evidence from US municipalities. Am. Econ. J. Econ. Policy **3**, 82–112 (2011)
27. Ministry of the Interior and Safety: Total and net budget. https://lofin.mois.go.kr
28. Park, J., Keum, J.: Analysis on the institutional, procedural and behavioral constraint of the citizen participatory budget system (CPBS): focused on the case of CPBS of Seoul Metropolitan Government (SMG). Korea J. Local Publ. Finance. **27**, 103–139 (2022)
29. Lee, S.K.: Policy formulation and implementation on participatory budgeting in Seoul, South Korea. Policy Gov. Rev. **1**, 125–137 (2017)
30. Seoul Metropolitan Government: Seoul metropolitan city participatory budgeting management ordinance (2022)
31. Seoul Metropolitan Government: 2021 Participatory Budgeting White Paper. Seoul Metropolitan Government (2021)
32. Shybalkina, I.: Toward a positive theory of public participation in government: variations in New York City's participatory budgeting. Publ. Adm. **100**, 841–858 (2022)
33. Sroska, J., Pawlica, B., Ufel, W.: Evolution of the civic budget in Poland – towards deliberation or plebiscite? Libron (2022)
34. Roberts, M.E., Stewart, B.M., Tingley, D.: stm: R package for structural topic models. J. Stat. Softw. **91**, 1–40 (2019)
35. Krippendorff, K.: Content Analysis: An Introduction to Its Methodology. Sage Publications, Thousand Oaks (2004)
36. Schmiedel, T., Müller, O., Vom Brocke, J.: Topic modeling as a strategy of inquiry in organizational research: a tutorial with an application example on organizational culture. Organ. Res. Methods. **22**, 941–968 (2019)
37. No, W., Hsueh, L.: How a participatory process with inclusive structural design allocates resources toward poor neighborhoods: the case of participatory budgeting in Seoul, South Korea. Int. Rev. Adm. Sci. **88**, 663–681 (2022)
38. Seoul Metropolitan Government: Operation of citizen participation budget. Seoul Metropolitan Government (2022)
39. No, W.: Mandated participatory budgeting in South Korea: issues and challenges. In: No, W., Brennan, A., Schugurensky, D., (eds.) By the People: Participatory Democracy, Civic Engagement, and Citizenship Education, pp. 109–117. Arizona State University (2017)
40. Seoul Metropolitan Government: Survey on Urban Policy Index (2021)
41. Ganuza, E., Baiocchi, G.: The power of ambiguity: how participatory budgeting travels the globe. J. Deliberative Democracy **8**, 1–12 (2012)

Digital Transformation

Barriers to the Introduction of Artificial Intelligence to Support Communication Experts in Media and the Public Sector to Combat Fake News and Misinformation

Walter Seböck⬚, Bettina Biron⬚, and Thomas J. Lampoltshammer(✉)⬚

Department for E-Governance and Administration, University for Continuing Education Krems,
Dr.-Karl-Dorrek-Str. 30, 3500 Krems an der Donau, Austria
`{walter.seboeck,bettina.biron,`
`thomas.lampoltshammer}@donau-uni.ac.at`

Abstract. Public trust represents a cornerstone of today's democracies, their media, and institutions and in the search for consensus among different actors. However, the deliberate and non-deliberate spreading of misinformation and fake news severely damages the cohesion of our societies. This effect is intensified by the ease and speed of information creation and distribution that today's social media offers. In addition, the current state-of-the-art for artificial intelligence available to everybody at their fingertips to create ultra-realistic fake multimedia news is unprecedented. This situation challenges professionals within the communication sphere, i.e., media professionals and public servants, to counter this flood of misinformation. While these professionals can also use artificial intelligence to combat fake news, introducing this technology into the working environment and work processes often meets a wide variety of resistance. Hence, this paper investigates what barriers but also chances these communication experts identify from their professional point of view. For this purpose, we have conducted a quantitative study with more than 100 participants, including journalists, press officers, experts from different ministries, and scientists. We analyzed the results with a particular focus on the types of fake news and in which capacity they were encountered, the experts' general attitude towards artificial intelligence, as well as the perceived most pressing barriers concerning its use. The results are then discussed, and propositions are made concerning actions for the most pressing issues with a broad societal impact.

Keywords: Fake News · Artificial Intelligence · Media Forensic · Journalism · Social Media · Public Sector

1 Introduction

The advent of the Internet has fundamentally changed how information is spread and perceived within societies. Not only are the entrance barriers much lower than in classical media, but the speed at which information can be shared worldwide is unrivaled.

© IFIP International Federation for Information Processing 2023
Published by Springer Nature Switzerland AG 2023
N. Edelmann et al. (Eds.): ePart 2023, LNCS 14153, pp. 67–81, 2023.
https://doi.org/10.1007/978-3-031-41617-0_5

The "post-factual" [1], also called "post-truth" [2], society is amid an "information war" and poses immense challenges for the media and the public sector within democracies [3, 4]. During the early stage of this revolution of interpersonal and mass communication via digital technologies, this paradigm change was perceived as a huge chance to reduce inequality by providing increased access to the public discourse and hence give a voice to virtually everybody, which in turn would ultimately support democracy within our societies [5, 6]. An assessment that still holds today. But the downside is that the easier access opens the door to disinformation from various (dangerous) sources [7]. In an age of innovation through knowledge for a sustainable, cohesive society [8], misinformation and fake news have a direct negative impact on public value creation through falsified or misleading information [9]. In this context, media [10] and public administrations [11] have a shared responsibility as gatekeepers to ensure the accuracy of public information. Due to this shared responsibility, we decided to focus our study on both parties from a combined point of view.

Following the argument of shared responsibility, journalists and the public sector are in a difficult situation. Trying to resolve misinformation and inform the public often results in the original misinformation being distributed even more intensively. This circumstance is partly due to the backfire effect [12]. This effect relates to potential cognitive biases within individuals and will cause feelings in cases the deepest beliefs or world views are "violated" by information that would contradict them. Consequently, the affected individuals will try to protect their beliefs even more vehemently and hence, render entirely the original intention of correction counterproductive. Also, studies have demonstrated that negative news is often more likely to be picked up and spread among the general public than positive news [13, 14].

Thus, the media and the public sector are in a problematic discrepancy between protecting free expression and disseminating information versus distorting democratic elections through massive disinformation campaigns. Moreover, in this tension range, they must deal with distrust and attacks often determined by prejudice, fear, and hate [15, 16]. This problematic situation is additionally pushed by social bots, which can massively spread whole global disinformation campaigns [17–19]. In addition, continuous development in artificial intelligence (AI), especially deep fakes, makes it increasingly challenging, even for experienced communication experts, to distinguish information from disinformation [20, 21].

But AI can also be a potent solution for identifying and fighting fake news. However, many barriers impede the implementation and use of tools by the leading media and the public sector to detect disinformation [20, 21]. To get a deeper understanding of those barriers, we conducted a quantitative survey with more than 100 experts from the field of leading media and the public sector, with a particular focus on the use of AI to fight disinformation.

The remainder of this paper is structured as follows: Sect. 2 provides a short discourse about state-of-the-art solutions using AI to combat fake news. In Sect. 3, we present the underlying methodology of this study and the collected data, including an overall profile of the participants. Section 4 then continues with the analysis of the results of the survey. After that, Sect. 5 discusses key learnings and practical implications. Section 6 then closes the paper with the conclusions and outlook for future work.

2 Related Work

A growing body of literature exists concerning technical solutions for using AI to combat fake news and misinformation. In this section, we provide a short discourse along the work of Shahid et al. [22] to inform the reader about state-of-the-art solutions currently used within available tools to the media and the public sector. Based on their analysis, current research streams can be separated into the following categories (ibid.):

- Automatic detection: the idea behind this approach is to extract features of fake news within deep learning models to be used for the automated classification of news items. Examples of this approach include the research of Ozbay and Alatas [22], who developed a solution to detect fake news in social media via a transfer process of unstructured data toward structured data, combined with a multi-algorithm analysis.
- Language-specific detection: this approach targets the development of a language-specific model beyond the limitation of English as the primary language. Studies that have used this approach, including the work of Faustini and Covões [23], build upon textual features and are not bound to a specific language, significantly increasing the overall usability, especially in an international context.
- Dataset-based detection: the main goal is to develop highly specialized datasets to test and challenge existing and newly developed algorithms. Examples include Neves et al. [24], who developed a method of removing fingerprints of algorithms (i.e., Generative Adversarial Networks) in face manipulation of images to challenge existing detection tools.
- Early detection: focuses on detecting fake news to limit its propagation at the earliest stage possible. Studies following this direction include Zhou et al. [25], who targeted the prevention of spreading fake news on social media via a supervised classification approach, building on social sciences and psychology theories.
- Stance detection: the idea behind this approach is not only to detect fake news but to deepen the underlying understanding of it. This is achieved by also including the stance of the reporting news outlets toward the reported event or incident. Research following this idea includes the work of Xu et al. [26], who integrated the reputational factors of news distributors, such as registration behavior, timing, ranking of domains, and their popularity.
- Feature-based detection: while this approach is similar to the automatic detection described before, it goes beyond classical textual features and includes topological and semantical features to improve the overall classification. Studies that have followed this idea include de Oliveira et al. [27], who incorporated stylistic information of social media posts, i.e., tweets, to improve the accuracy of fake news detection.
- Ensemble learning: the concept behind this approach is to use not one but a combination of multiple algorithms to identify and classify fake news. Examples of such combined approaches include Elhadad et al. [28], who addressed the issue of misleading information in the context of the COVID-19 pandemic, combining ten machine-learning algorithms with several feature extraction approaches.

3 Methodology and Data

In order to derive recommendations on how AI tools can be used for disinformation detection for leading media and the public sector, it is crucial to consider several factors, motivations, and potential barriers. These include challenges with implementation and the working environment, technological maturity, data protection, uncertainty about AI, and advancing technological progress in general. To address this challenging domain rigorously, a questionnaire was created during the applied research project *defalsif-AI (Detection of Disinformation via Artificial Intelligence)* aimed at communication experts. The questionnaire was created based on literature around dimensions of fake news, misinformation, and information disorder [5, 29, 30], with a particular focus on professionals and their perspectives on i) the types of media to be confronted with, ii) individual detection approaches, iii) types of fake news encountered, iv) attitudes toward AI technological progress, as well as v) experience on currently used AI tools in the respective working environments.

In the first step, this questionnaire was circulated among the consortium partners, and in the second step, a snowball-based system was applied to other related areas. This approach helped us to significantly increase the overall coverage of experts that deal with fake news within their professional environment.

These experts included journalists, press officers, experts from different ministries, and scientists dealing with the topic of fighting fake news and disinformation. Overall, we collected n = 106 completed surveys. Since the population in these sectors is unknown, it is seldom possible for expert surveys to be representative. Nevertheless, they allow profound assessments to be made of trends among professionals.

In addition to demographic data and questions about media genres and usage behavior, the questionnaire focused on the frequency and risk of dealing with fake news and misinformation in everyday work, intuitive and technological detection, research activities, and, last but not least, the desire for or possible rejection of AI-based software. The survey was conducted from May to June 2021.

49% of the respondents work in the media domain (journalists, press officers, PR professionals, etc.), while 31% in communication and security, including fields such as the police or the Ministry of the Interior. 7% of the respondents work in the field of diplomacy or the Ministry of Foreign Affairs, and 13% in the field of research.

Concerning the age distribution of our participants, about 85% of them resided within the mid-career and late-career levels.

Regarding professional experience, about 62% had more than ten years of professional experience. Hence, a high level of insight and proficiency is represented among the study participants.

4 Analysis and Results

4.1 Fake News and Misinformation Within Working Environments

Only 23.6% of the respondents see little or no threat to democracy in disinformation, and 76.4% of the respondents consider fake news to be a high or very high risk for democracy. In the context of AI-based media forensics, it is necessary to understand the medium through which experts often come into contact with disinformation (see Fig. 1).

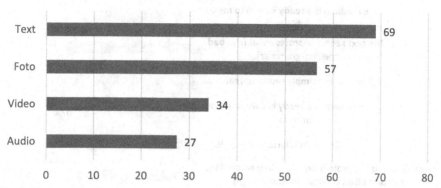

Fig. 1. Types of Mediation (n = 106); agreement high and very high (in %, n = 106; 6-point Likert scale; multiple answers possible).

Most subjects are confronted with disinformation via text, followed by manipulated photos often or very often. Text and photos are also the favored means of communication in traditional media, although video and audio are becoming increasingly popular, primarily through social media. In this context, this also raises the question of whether even experts can recognize manipulation, given the rapid technological development of deep fakes by video and audio. Studies also indicate that time of day, emotional state, fatigue, or age can significantly detect deepfakes [31, 32]. Concerning the odds of sharing misinformation, such as deep fakes, between individuals with a high interest in politics and those without, they later seem more prone to forwarding such misinformation [33]. In addition, personality traits such as optimism, especially for social media, can also play a role in classifying and spreading [34]. Ahmed points out that there is still limited knowledge about how social media users deal with this newer form of disinformation [33]. Our survey reveals a similar picture asking about the experts' strategies (see Fig. 2) in case of suspicion of fake news, and the following picture emerges.

Research whether and how other media report on it (78.5%), a critical look at the imprint of the medium (64,5%); checking the background of the author (54,2%), research how coherent contextual information such as geographic data, weather data, etc. are (39,3%), using fact-checking services like Mimikama, Correctiv, Hoaxsearch, etc. (28,9%), reverse image search on the Internet to check the actual origin of an image like Reverse Google Image Research, tineye.com Yandex (24.3%), and checking the metadata of an image (14,9%).

Since technology is advancing increasingly in mass manipulation, the results could indicate that training and AI-based tools will be increasingly necessary, especially for detecting deep fakes [35]. Especially since the sinister combination of manipulated videos and WhatsApp, e.g., in India [36], has already led to lynching mobs with innocent deaths, a change of modalities and further studies, training, and detection tools seem to be necessary for the context of security.

Continuing our analysis, we asked the participants to name the type(s) of misinformation and fake news most relevant to them in their daily business (see Fig. 3).

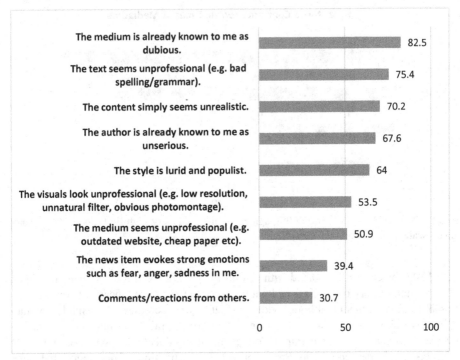

Fig. 2. Intuitive detection: Question: Based on which indications do you intuitively suspect whether it could be Fake News? (n = 106; agreement high and very high (in %, n = 106; 6-point Likert scale; multiple answers possible)).

The respondents stated that they were mainly involved in news fabrication in their professional life. Fabrication in this context implies that the generated news items are not based on facts. However, due to their style and presentation, they create the impression with readers that they are real. Similar to fabricated news is propaganda, usually originating from a political motivation to either praise or discredit an individual or entity. Examples of such approaches, besides others, can be found within official Russian news channels, deliberately using narratives to convey a particular image to their audience [3]. Similarly, tear-jerking misleading headlines were used to create click-bait and were frequently named by our respondents as a challenge they have to cope with within their own professional routine.

In the second place, however, are already photo and video manipulation. This observation is only, at first glance, contradictory to Fig. 1, in which photo and video manipulation are not classified as particularly frequent. It seems reasonable to assume that these manipulations are challenging to recognize precisely because of the technical know-how and effort; thus, the motives behind them must be exceptionally high. The respondents least frequently mentioned mouse-to-mouse propaganda, i.e., paid customer reviews, which are popular with large online retailers.

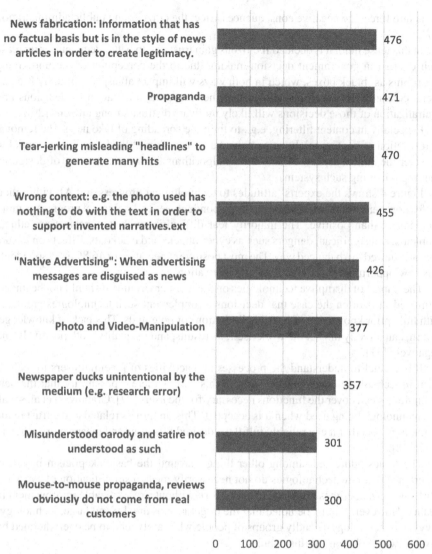

Fig. 3. Question: Which types of fake news are particularly relevant to you professionally? (n = 106; agreement high and very high (in %, n = 106; 6-point Likert scale; multiple answers possible)).

4.2 Barriers and Trust in AI

The application of technologies in the context of decision-making in the public sector always impacts the lives of citizens. Reasons for the introduction of these technologies often include cost-saving, increased efficiency, and improved 'objectivity' due to 'fair' algorithms [27]. Yet these technologies can also trigger unintended side-effects, which bear risks that are hard to foresee, measure, and thus be prepared for. When these risks come into force, the negative consequences affect the citizens and public administration [27].

In this tense field, it is decided if citizens gain trust due to better decisions or increase their distrust in government decision-making due to the perception of the underlying algorithms as 'black boxes, which in both ways will impact all aspects of daily life and social cohesion. Research has shown that the increased automation of decisions and centralization of those decisions will likely motivate distrust among citizens [29].

Especially in content filtering, e.g., to fight the spreading of fake news, the removal or restriction of content might be perceived as censorship [30]. Hence, it is essential to consider the ethical aspects of data-driven algorithms from the beginning of designing and implementing such systems.

Figure 4 shows the experts' attitudes to technological progress and AI within their professional environment. In general, the respondents see technological progress as more problematic than positive. The majority fear difficulties with data protection, ethical problems, and significant dangers such as cyber-attacks and blackouts. Effects on leisure time are viewed in a balanced way. The most positive expectation of 46.8% of respondents was new opportunities for creativity and innovation.

The impact of disruptive technologies on the work environment should not be underestimated. It is often the case that decisions to implement such technologies are made with little prior knowledge of possible limitations or potentials. This lack of knowledge, in turn, can directly impact the work itself, its results, and its quality, both positively and negatively [37].

It is crucial to understand the processes and activities of potential users to include them in technology development. Only in this way is there a possibility that the new technologies can cover the functions necessary for the users [37]. Grabowski et al. speak of a technology being used when it is accepted. This, in turn, is related to the trust in the technology, whether it can reliably fulfill the desired functions and means more efficient work [38].

The topics addressed, among other things, around the basic skepticism based on experience that new technologies do not necessarily mean a simplification of everyday work but can sometimes even lead to more work without a recognizable improvement in quality. However, it must be noted that the target groups are, by and large, technology-savvy and technology-friendly groups of people who rarely tend to be overwhelmed by new software solutions in this context.

Turning to our last part of the survey, we asked the participants to express their opinion concerning barriers to using a fake news detection software tool in their workplace (see Fig. 5).

In addition to a lack of application options, the respondents see unclear or non-transparent strategies, high time and cost expenditure, and a lack of customized solutions as obstacles to using software solutions for fake news detection. Lack of acceptance by the workforce and high demands on data protection and security is mentioned the least, but more than a third of the respondents still cite them as possible obstacles. Winning the acceptance of employees should therefore be considered in training courses.

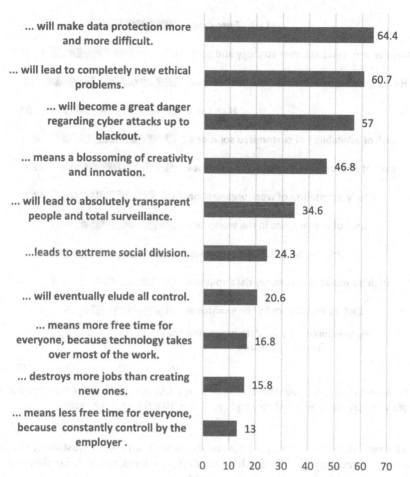

Fig. 4. Attitudes to technological progress and AI; agreement high and very high (in %, n = 106; 6-point Likert scale).

5   Lessons Learned and Propositions

The analysis of our survey has demonstrated the most pressing barriers that experts from the media and the public sector currently see in using AI to fight fake news and disinformation. Amongst the top-ranking results were: i) lack of trust in the technology, ii) in-transparent organizational strategies, and iii) ethical and privacy concerns. Hence, in the following, we provide some selected propositions and discussion points of lessons learned and what needs to be addressed to overcome the identified barriers.

In tools and data, we trust – attitudes towards AI as a 'Colleague'. Using AI to identify and communicate fake news to the general public is not without criticism, and trust in the technology is one of the key issues to ensure acceptance [39]. The literature shows that the same norms often come into play here as in interpersonal interaction [40]. In this context, it is also essential to consider that people tend to perceive AI as a

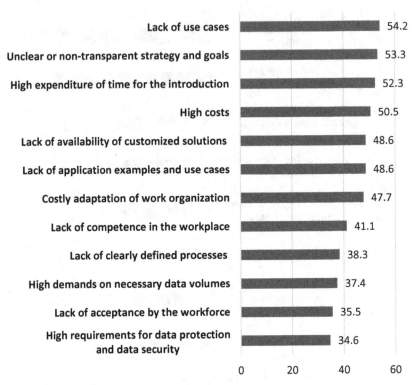

Fig. 5. What do you think would be barriers to using a fake news detection software tool in your workplace? (Agreement high and very high; in %, n = 106; 6-point Likert scale).

"counterpart" and not as a tool [41]. AI and its results must also be trustworthy in times of personal uncertainty [42]. In-depth research into the influence of perception and trust in the context of AI is, therefore, necessary [43].

I know it as well as the back of my hand – the importance of personal experience with AI. Many users have considerable reservations about AI-based fact-checking tools [44]. Overcoming these reservations is an open challenge due to such tools' increasing distribution and use [45]. The accuracy of the analysis results is not always the decisive aspect of whether users trust the tools [43]. The users' understanding of how to use the tools and how they work can have a lasting influence on their trust in the technology [46]. Personal experience in dealing with these tools [47] can also lead to realistic expectations of the tool itself [48] and, thus, to a more positive attitude toward AI [49]. It is, therefore, essential to define solutions that embed the presentation of results and the handling of the AI tool in the user's experience. If this succeeds, it could lead to greater self-reflection and a more critical approach to news and information through fact-checkers and evaluation tools [43].

Digital ethics – the importance of societal consensus and consent. Following the paradigm of digital humanism as a mindset of understanding the highly entangled and complex relationship of humans and technology [50], ultimately, technology should foster the free development of the individual to their full potential, but at the same time, not negatively impact others. This view also implies that tendencies towards anti-humanism through technology, e.g., artificial intelligence, should be identified and questioned [51]. This demand necessitates the fundamental need for ethical considerations embedded in all organizational processes. The essential question at hand: where to start? A plethora of frameworks is targeted at the ethical aspects of AI, where interested individuals can quickly lose oversight [52].

Furthermore, many of these frameworks are either on a high meta-level and thus hard to operationalize or on the opposite side, i.e., specific for a particular field or domain; hence, transferability is often limited [53]. Consequently, an approach needs to be selected that allows experts in communication to map common principles of digital ethics and the use of AI into their domain. Becker et al. have developed a three-step approach, i.e., analysis of principles, mapping the derived principles, and deriving an individual code of digital ethics [54]. Adopting this or similar frameworks can support communication experts in building their respective codes of conduct and guidelines for using AI. This adoption would ease internal barriers, as most refer to the missing knowledge and transparent and understandable guidelines.

6 Conclusions

Our study among the professionals has demonstrated that the situation is critical and that although AI can be a significant support within the daily work of communication experts, it is a blessing and a curse simultaneously. While the technology enables them to identify potentially fake news and misinformation, they struggle to communicate the results quickly and reach the necessary target audience. They are also facing fears and rejection concerning the use of AI by the general public. Censorship, violation of the free press, and intended overblocking are only a selection of accusations they are confronted with. This backlash leads to the build-up of internal barriers to adopting artificial intelligence within their organization. One of the biggest challenges comes in the lack of internal knowledge and capacity, which is also reflected in many follow-up barriers, such as fear of data privacy violation, mass surveillance, societal dived, or personal liability. What would be required is sophisticated training and proper adaptions to existing processes and work routines.

Consequently, this would lead to a deeper understanding of the underlying technology, its capabilities, and its limitations. In this context, the transparency of the use of algorithms and tools and the underlying decision process of these tools would be increased. Consequently, the responsible use would be strengthened, as well as the overall accountability for the application, interpretation, and dissemination of results. This overall increased knowledge would also become beneficial in terms of privacy protection while working with various sources of data and information.

For future work, several paths opened up based on our study results. The discussion around the regulation of AI within the EU is currently omnipresent. Thus, an examination of to what extent the handling of disinformation is regulated on a national level in the DACH countries and on an EU level (e.g., GDPR, Digital Services Act) or which initiatives exist in this regard in order to develop a well-founded recommendation for the future regulation of disinformation will be of interest. For a responsible approach to AI-based disinformation detection, the significance of the EU's AI Act is of high importance to the research community and the community of practitioners, and also, what consequences are to be drawn from a legal perspective. In addition to the provisions of the AI Act, national developments should also be considered to develop a framework for the legal, ethical, and transparent use of AI systems to detect disinformation. The aim is to shed light on the legal framework for designing AI systems to detect disinformation and to make recommendations based on a comprehensive consideration of the fundamental rights of the citizens affected.

Another interesting aspect for future research comes from the ever-increasing flood of disinformation, not least multiplied by bots, trolls, and generative AI, which raises concerns about the destabilization of society and a post-factual future. Technological development enables the massive increase of disinformation in quantity and quality while, at the same time, also providing solutions in the area of detection. However, paying particular attention to this ambivalent relationship to AI is vital, especially in the context of information dissemination in society. A representative survey of the Austrian population will empirically record the rejection, fears, and hopes regarding various aspects such as data protection, freedom of opinion, "overblocking" and transparency. From this data material, concrete recommendations for action are derived from promoting the acceptance of a broad population and taking ethics and diversity into special consideration.

Acknowledgments. The work described in this paper was funded in the context of the defalsif-AI project (FFG project number 879670, funded by the Austrian security research program KIRAS of the Federal Ministry of Agriculture, Regions, and Tourism BMLRT).

References

1. Hossová, M.: Fake news and disinformation: phenomenons of post-factual society. Media Literacy Acad. Res. **1**, 27–35 (2018)
2. Bybee, C.: Can democracy survive in the post-factual age?: A return to the Lippmann-Dewey debate about the politics of news. Journal. Commun. Monographs **1**, 28–66 (1999)
3. Khaldarova, I., Pantti, M.: Fake news: the narrative battle over the Ukrainian conflict. Journal. Pract. **10**, 891–901 (2016). https://doi.org/10.1080/17512786.2016.1163237
4. Seboeck, W., Biron, B., Lampoltshammer, T.J., Scheichenbauer, H., Tschohl, C., Seidl, L.: Disinformation and fake news. In: Masys, A.J. (ed.) Handbook of Security Science, pp. 1–22. Springer, Cham (2020). https://doi.org/10.1007/978-3-319-51761-2_3-1
5. Fraas, C., Klemm, M., Gesellschaft für Angewandte Linguistik (eds.) Mediendiskurse: Bestandsaufnahme und Perspektiven. P. Lang, Frankfurt am Main ; New York (2005)
6. Kriesi, H., Lavenex, S., Esser, F., Matthes, J., Bühlmann, M., Bochsler, D.: Democracy in the Age of Globalization and Mediatization. Palgrave Macmillan UK, London (2013). https://doi.org/10.1057/9781137299871
7. Bennett, W.L., Livingston, S.: The disinformation order: disruptive communication and the decline of democratic institutions. Eur. J. Commun. **33**, 122–139 (2018). https://doi.org/10.1177/0267323118760317

8. Carayannis, E.G., Barth, T.D., Campbell, D.F.: The Quintuple Helix innovation model: global warming as a challenge and driver for innovation. J. Innov. Entrepreneurship. **1**, 1–12 (2012)
9. Van Meter, H.J.: Revising the DIKW pyramid and the real relationship between data, information, knowledge, and wisdom. Law Technol. Hum. **2**, 69–80 (2020)
10. Guo, L.: China's "fake news" problem: exploring the spread of online rumors in the government-controlled news media. Digit. Journal. **8**, 992–1010 (2020)
11. Ninkov, I.: Separating truth from fiction: legal aspects of "fake news." Biztonságtudományi Szemle. **2**, 51–64 (2020)
12. Wood, T.J., Porter, E.: The elusive backfire effect: mass attitude' steadfast factual adherence. Polit. Behav. **41**, 135–163 (2019)
13. Huijstee, D., Vermeulen, I., Kerkhof, P., Droog, E.: Continued influence of misinformation in times of COVID-19. Int. J. Psychol. ijop.12805 (2021). https://doi.org/10.1002/ijop.12805
14. Jacobson, N.G., Thacker, I., Sinatra, G.M.: Here's hoping it's not just text structure: the role of emotions in knowledge revision and the backfire effect. Discourse Process. 1–23 (2021). https://doi.org/10.1080/0163853X.2021.1925059
15. Appel, M. (ed.): Die Psychologie des Postfaktischen: über Fake News, "Lügenpresse" Clickbait & Co. Springer, Heidelberg (2020). https://doi.org/10.1007/978-3-662-58695-2
16. Hagen, L.: Nachrichtenjournalismus in der Vertrauenskrise. "Lügenpresse" wissenschaftlich betrachtet: Journalismus zwischen Ressourcenkrise und entfesseltem Publikum. ComSoz. **48**, 152–163 (2015). https://doi.org/10.5771/0010-3497-2015-2-152
17. Hajli, N., Saeed, U., Tajvidi, M., Shirazi, F.: Social bots and the spread of disinformation in social media: the challenges of artificial intelligence. Brit. J. Manag. 1467–8551.12554 (2021). https://doi.org/10.1111/1467-8551.12554
18. Shao, C., Ciampaglia, G.L., Varol, O., Flammini, A., Menczer, F.: The spread of fake news by social bots. **96**, 104. arXiv preprint arXiv:1707.07592 (2017)
19. Wang, P., Angarita, R., Renna, I.: Is this the era of misinformation yet: combining social bots and fake news to deceive the masses. Presented at the Companion Proceedings of the Web Conference 2018 (2018)
20. Zhang, T.: Deepfake generation and detection, a survey. Multimedia Tools Appl. **81**, 6259–6276 (2021). https://doi.org/10.1007/s11042-021-11733-y
21. Mirsky, Y., Lee, W.: The creation and detection of deepfakes: a survey. ACM Comput. Surv. **54**, 1–41 (2022). https://doi.org/10.1145/3425780
22. Ozbay, F.A., Alatas, B.: Fake news detection within online social media using supervised artificial intelligence algorithms. Physica A: Stat. Mech. Appl. **540**, 123174 (2020)
23. Faustini, P.H.A., Covoes, T.F.: Fake news detection in multiple platforms and languages. Expert Syst. Appl. **158**, 113503 (2020)
24. Neves, J.C., Tolosana, R., Vera-Rodriguez, R., Lopes, V., Proença, H., Fierrez, J.: Ganprintr: improved fakes and evaluation of the state of the art in face manipulation detection. IEEE J. Sel. Top. Sig. Process. **14**, 1038–1048 (2020)
25. Zhou, X., Jain, A., Phoha, V.V., Zafarani, R.: Fake news early detection: a theory-driven model. Digit. Threats Res. Pract. **1**, 1–25 (2020)
26. Xu, K., Wang, F., Wang, H., Yang, B.: Detecting fake news over online social media via domain reputations and content understanding. Tsinghua Sci. Technol. **25**, 20–27 (2019)
27. de Oliveira, N.R., Medeiros, D.S., Mattos, D.M.: A sensitive stylistic approach to identify fake news on social networking. IEEE Sig. Process. Lett. **27**, 1250–1254 (2020)
28. Elhadad, M.K., Li, K.F., Gebali, F.: Detecting misleading information on COVID-19. IEEE Access **8**, 165201–165215 (2020)
29. Allcott, H., Gentzkow, M.: Social media and fake news in the 2016 election. J. Econ. Perspect. **31**, 211–236 (2017). https://doi.org/10.1257/jep.31.2.211
30. Wardle, C., Derakhshan, H.: Information disorder: toward an interdisciplinary framework for research and policymaking. Council of Europe Strasbourg (2017)

31. Jung, T., Kim, S., Kim, K.: Deepvision: deepfakes detection using human eye blinking pattern. IEEE Access **8**, 83144–83154 (2020)
32. Müller, N.M., Pizzi, K., Williams, J.: Human perception of audio deepfakes. Presented at the Proceedings of the 1st International Workshop on Deepfake Detection for Audio Multimedia (2022)
33. Ahmed, S.: Who inadvertently shares deepfakes? Analyzing the role of political interest, cognitive ability, and social network size. Telematics Inform. **57**, 101508 (2021)
34. Valenzuela, S., Halpern, D., Katz, J.E., Miranda, J.P.: The paradox of participation versus misinformation: social media, political engagement, and the spread of misinformation. Digit. Journal. **7**, 802–823 (2019). https://doi.org/10.1080/21670811.2019.1623701
35. Weerawardana, M., Fernando, T.: Deepfakes detection methods: a literature survey. In: 2021 10th International Conference on Information and Automation for Sustainability (ICIAfS), pp. 76–81 (2021). https://doi.org/10.1109/ICIAfS52090.2021.9606067
36. Sundar, S.S., Molina, M.D., Cho, E.: Seeing is believing: is video modality more powerful in spreading fake news via online messaging apps? J. Comput.-Mediat. Commun. **26**, 301–319 (2021). https://doi.org/10.1093/jcmc/zmab010
37. Pennathur, P.R., Bisantz, A.M., Fairbanks, R.J., Perry, S.J., Zwemer, F., Wears, R.L.: Assessing the impact of computerization on work practice: information technology in emergency departments. In: Proceedings of the Human Factors and Ergonomics Society Annual Meeting, vol. 51, pp. 377–381 (2007). https://doi.org/10.1177/154193120705100448
38. Grabowski, M., Rowen, A., Rancy, J.-P.: Evaluation of wearable immersive augmented reality technology in safety-critical systems. Saf. Sci. **103**, 23–32 (2018). https://doi.org/10.1016/j.ssci.2017.11.013
39. Gillath, O., Ai, T., Branicky, M.S., Keshmiri, S., Davison, R.B., Spaulding, R.: Attachment and trust in artificial intelligence. Comput. Hum. Behav. **115**, 106607 (2021). https://doi.org/10.1016/j.chb.2020.106607
40. Nass, C., Moon, Y.: Machines and mindlessness: social responses to computers. J. Soc. Isssues **56**, 81–103 (2000). https://doi.org/10.1111/0022-4537.00153
41. Seeber, I., et al.: Machines as teammates: a research agenda on AI in team collaboration. Inf. Manag. **57**, 103174 (2020). https://doi.org/10.1016/j.im.2019.103174
42. Okamura, K., Yamada, S.: Adaptive trust calibration for human-AI collaboration. PLoS ONE **15**, e0229132 (2020). https://doi.org/10.1371/journal.pone.0229132
43. Shin, J., Chan-Olmsted, S.: User perceptions and trust of explainable machine learning fake news detectors. Int. J. Commun. **17**, 23 (2022)
44. Brandtzaeg, P.B., Følstad, A.: Trust and distrust in online fact-checking services. Commun. ACM. **60**, 65–71 (2017). https://doi.org/10.1145/3122803
45. Zhou, X., Zafarani, R.: A survey of fake news: fundamental theories, detection methods, and opportunities. ACM Comput. Surv. **53**, 1–40 (2021). https://doi.org/10.1145/3395046
46. Siau, K., Wang, W.: Building trust in artificial intelligence, machine learning, and robotics. Cutter Bus. Technol. J. **31**, 47–53 (2018)
47. Mohseni, S., Zarei, N., Ragan, E.D.: A Multidisciplinary survey and framework for design and evaluation of explainable AI systems. ACM Trans. Interact. Intell. Syst. **11**, 1–45 (2021). https://doi.org/10.1145/3387166
48. Matthews, G., Lin, J., Panganiban, A.R., Long, M.D.: Individual differences in trust in autonomous robots: implications for transparency. IEEE Trans. Human-Mach. Syst. **50**, 234–244 (2020). https://doi.org/10.1109/THMS.2019.2947592
49. Araujo, T., Helberger, N., Kruikemeier, S., de Vreese, C.H.: In AI we trust? Perceptions about automated decision-making by artificial intelligence. AI Soc. **35**(3), 611–623 (2020). https://doi.org/10.1007/s00146-019-00931-w

50. Hofkirchner, W., Kreowski, H.-J.: Digital humanism: how to shape digitalisation in the age of global challenges? In: IS4SI 2021, p. 4. MDPI (2022). https://doi.org/10.3390/proceedings2022081004

51. Schmölz, A.: Die Conditio Humana im digitalen Zeitalter: Zur Grundlegung des Digitalen Humanismus und des Wiener Manifests. MedienPädagogik. 208–234 (2020). https://doi.org/10.21240/mpaed/00/2020.11.13.X

52. Floridi, L., Cowls, J.: A unified framework of five principles for AI in society. Harvard Data Sci. Rev. (2019). https://doi.org/10.1162/99608f92.8cd550d1

53. Hickok, M.: Lessons learned from AI ethics principles for future actions. AI Ethics 1(1), 41–47 (2020). https://doi.org/10.1007/s43681-020-00008-1

54. Becker, S.J., Nemat, A.T., Lucas, S., Heinitz, R.M., Klevesath, M., Charton, J.E.: A code of digital ethics: laying the foundation for digital ethics in a science and technology company. AI Soc. (2022). https://doi.org/10.1007/s00146-021-01376-w

Structuring Continuous Education Offers for E-Government-Competence Acquisition: A Morphological Box

Holger Koelmann[1]([✉]) [iD], Michael Koddebusch[1] [iD], Julia Bücker[2] [iD],
Marc Egloffstein[3] [iD], and Jörg Becker[1] [iD]

[1] University of Münster - ERCIS, Münster, Germany
{koelmann,koddebusch,becker}@ercis.de
[2] University of Münster - Institute of Education, Münster, Germany
julia.buecker@uni-muenster.de
[3] University of Mannheim - Area of Economic and Business Education, Mannheim,
Germany
egloffstein@uni-mannheim.de

Abstract. Being involved in all facets of the digital transformation of the public sector, public officials play a pivotal role in successfully pursuing the transformative process. Hence, they must be equipped with the means to actively partake in creating a digitalized public sector, making the acquisition of e-government-competences indispensable. However, we must observe an increasing gap between the degree of required and obtained e-government-competences in the workforce of public sector organizations in developed countries, especially in examples such as Germany. This is also due to public officials being unable to select and attend the individually appropriate continuous education offers for them. To provide a means for structuring the decision-relevant criteria when choosing continuous education offers for e-government-competences, we develop a morphological box depicting the conceptual dimensions for such offers as the main result of this study. To that end, we conducted an in-depth interview study, aggregating the perspectives of relevant stakeholders from the German public sector.

Keywords: E-Competences · E-Government · Public Sector
Digitalization · Continuous Education

1 Introduction

The pursuit of digital transformation in the public sector faces many challenges and is progressing only stagnantly, even in some developed countries [51]. A key factor in advancing the digitalization of the public sector are its employees. Scholars have repeatedly argued for the importance of public officials in pursuing digital transformation initiatives in the public sector, being an integral part of bureaucratic activities [17,27,41]. Based on individual characteristics, the public

© IFIP International Federation for Information Processing 2023
Published by Springer Nature Switzerland AG 2023
N. Edelmann et al. (Eds.): ePart 2023, LNCS 14153, pp. 82–98, 2023.
https://doi.org/10.1007/978-3-031-41617-0_6

official is attributed with an ambivalent role as either a facilitator of or an impediment to the transformative process [1,8,36]. It is therefore important for public officials to acquire e-government-competences (in the following to be referred to as e-competences, which we define as public sector-specific digitalization competences [see Sect. 2]). However, despite the importance, many public officials have not yet successfully obtained a sufficient degree of e-competences, leading to a growing e-competence gap among the total public workforce [24]. Looking at the already considerable shortage, which is due to demographic developments [5] going to increase further, it will not suffice to recruit new staff with the required e-competences. Hence, it is imperative to provide targeted continuous education opportunities for the acquisition of e-competences to existing staff that they can seize on-the-job. Even though we can observe an increasing amount of scholars promoting the idea of reinforced e-competence acquisition [12,22,33] and practitioner-oriented research projects being set up for this purpose[1], existing means for continuous education do not seem to fill this gap fully [39]. Reasons for that are manifold: for example, offers are often not flexible enough to adapt to the learners' circumstances, such as their prior knowledge, work focus, and time availability. Another reason is the lack of structural means to assess continuous education offers in terms of their individual configuration. This, in turn, impedes the learners' ability to evaluate the adequacy of an offer for their own goals. As, over time, public organizations became increasingly aware of this issue, providers of continuous education offers started adopting ICT-enabled formats, such as Massive Open Online Courses (MOOCs) [20]. However, despite these efforts, the desired effect has not fully manifested. Even though MOOCs are a widely used approach to continuous education, giving learners time-, space-, and financial flexibility, they have high dropout rates [7] and often lower instructional quality than conventional formats [31], suggesting that their promise was too big for them to keep. However, since the advent of MOOCs we have seen some exciting developments aside from that. New formats, such as micro-learning, are enabled by digital means and have become increasingly popular by offering new avenues in teaching and learning [3]. The increasing amount and diversity of different offers, however, poses a challenge in itself: continuous education offers are often developed based on technological possibilities and neglect the heterogeneity of their audience. Consequently, public officials, who should be at the core of attention, cannot find appropriate continuous education offers for their individual needs. Therefore, we want to provide a structuring instrument that accounts for the delineation of decision-relevant criteria when choosing continuous education offers. Hence, our research goal is:

The construction of a morphological box depicting the conceptual dimensions and their respective characteristics of continuous education offers for e-competences.

This goal is achieved by aggregating the perspectives of public officials utilizing continuous education offers and those responsible for continuous education

---

[1] E.g., see Qualifica Digitalis or eGov-Campus.

offers in their respective organizations. Our results contribute to e-competence literature in particular and, thus, to e-government literature in general.

2 Research Background

2.1 E-Competence and Public Sector Digitalization

The relationship between digitalization and the public official is bidirectional: while we know that the public sector's workforce has an impact on its digital transformation process [1,8,17,36], digitalization itself also impacts the everyday work of public servants. Consequently, digitalization demands new competences and a digital mindset of public servants [34]. According to Weinert [52] competences in general can be defined as individual's available or learnable abilities and skills, which enable this individual to solve specific problems. They also include the ability and preparedness to use them successfully in different contexts. The motivation and the will to do so is inherent. For the public sector, we define the required competences - in line with other scholars [35,38] as *e-competences*: the combination of an *"individual's work-related knowledge, skills and abilities"* [12] *required to act in a digitalized public sector*.

In recent years, researchers made an effort to establish and classify relevant e-competences and classify them [12,21,22]. One of the core messages of these publications is that public officials need e-competences that *"go beyond pure ICT skills"* [21] to actively partake in the digital transformation. For this work, we draw on Hunnius et al. [22], who classify e-competences into five categories: (1) technical, (2) socio-technical, (3) organizational, (4) managerial, and (5) political-administrative e-competences, comprising 14 distinct e-competences. Once having acquired such e-competences, researchers attest that they empower public officials to participate in new ventures, e.g., concerning public service delivery [29], smart city projects [43], and digital-ready legislation [25]. However, despite the outlined evidence, we must observe that there is an increasing e-competence gap, of which public sector organizations and actors from the political sphere appear to be well aware but do not seem to have found the means to close it yet [18,24]. For example, Koddebusch et al. [24] have found that even though the awareness of the importance of e-competence has grown over the past years, qualified staff (within and outside the own organization) is becoming increasingly scarce. The rootcause for this circumstance goes back to the education of public servants: undergraduate education, graduate education, and apprenticeships for public officials do not typically prioritize e-competences and digitalization-related matters but still focus on legal, economic, and management-oriented aspects [19]. Hence, university courses and apprenticeships need to re-focus their programs; and while they slowly start doing so [47], and there are even first attempts to structure education for e-Government positions for multiple roles [23], this only addresses future public officials. Also, specific research for positions, such as the government chief information officer [14], exists, but still only focuses on the original education programs of these people. Thus, it does not help mitigate the current e-competence shortage, which is likely to increase, considering the demographic change and, consequently, the loss of organizational knowledge [5,26].

E-competences must be acquired by people on the job today, and continuous education must be at the core of educational efforts. Unfortunately, there is nearly no secured knowledge of European efforts to promote professional development and/or training of public officials about e-competences (or their synonyms). Neither the scientific literature nor the statistical office of the European Union provide comprehensive information. On the one hand, we can find publications that deal with the fundamental education (e.g., undergraduate/postgraduate programs and apprenticeships) of public officials [46]. On the other hand, some publications deal with non-European and very organization-specific matters of professional development, which can hardly be generalized or are outdated [11,45,48].

Solely a McKinsey study [50] refers to a survey performed among 165 leaders in public sector organizations. The results show that the German public sector does not offer sufficient means for professional development and continuous education for further competence acquisition. 14% of the respondents state that the lack of these possibilities is among the main reasons for public servants leaving for the private sector. The study lists a "continuous education offensive" as key to making the public sector more attractive as an employer.

Despite the lack of clear scientific evidence, the growing e-competence shortage indicates that current means of professional continuous education are either not sufficient or the training formats do not meet the target group's requirements. Hence, it is worth considering innovative education formats that provide new possibilities.

2.2   From (E-)Competences to Innovative Training Formats

The acquisition of competences is a complex process, which takes place parallel to the processes of providing knowledge and achieving progress in ability [28]. The acquisition of competences is embedded in a combination of teaching, learning, knowledge, and ability [28]. Competences can be developed within formal and informal learning activities [10]. Formal learning activities can be characterized by their curricular form: often, they are controlled from the outside and take place in environments specially arranged for that purpose [6]. Informal learning often happens spontaneously and outside of "formally-designated learning contexts" [6]. Furthermore, non-formal learning activities can be distinguished, which are embedded between formal and informal learning activities [37]. Organized learning activities can be created within different formats, e.g., offline during a seminar. Considering the ongoing digital transformation, new formats have emerged which can transfer learning and development of competences into the digital space.

Originating in academic education, MOOCs have become an established format for training and digital learning in the workplace [13], and are also used for competence development in the public sector [20]. MOOCs are openly accessible online learning environments where learners can enroll for free or at a low-cost [54]. From an instructional design perspective, MOOCs are video-based self-learning environments with additional materials, assignments, and, in some cases, cooperative or collaborative elements. Research has shown that the traditional course-like MOOC format faces some challenges, such as high dropout rates [7] or a tendency toward low instructional quality [31]. Blending MOOCs and traditional instruction in professional training has successfully addressed these challenges [30]. Likewise, micro-formats such as Mini-MOOCs [49] or Learning Nuggets - instructional formats "primarily comprised of tasks that learners will undertake in a particular context to attain specific learning outcomes" [2] can provide an alternate pathway to competence development. Microlearning, commonly defined by bite-sized lessons, is considered an effective model for professional learning [53], which offers great potential for combining video-based MOOCs and mobile learning [3]. Micro-credentials, awarded after such short learning experiences and mostly tied to the achievement of a specific skill, are a means of credentialing that often better meets the demands of training and professional learning than academic certificates [16].

3   Research Method

To gain the empirical insights necessary to discover the important configuration dimensions of considered continuous education formats, we conducted semi-structured interviews with participants from different parts of the German public sector. During this, we followed the approach by Brinkmann [4] and Rowley [44]. The semi-structured interview format allowed us to change our questions and their order during the interview depending on respondents' answers and ask follow-up questions about new topics of interest raised by respondents [4,44].

To address all aspects of our research question, we developed a structured interview guideline covering the four main topics informed by the described findings of previous research:

1. *Professional Background and Context*: The interviewees professional background as well as experience with and potential responsibility for continuous education formats in their work environment in order to be able to place the statements of the interviewees in their respective work context.
2. *Status Quo in the Organization*: Status quo and potential issues with the current landscape of continuous education formats in their work environment to grasp the existing offers available to the interviewee, since existing continuous education formats are often found to be lacking in practice [50].
3. *Innovative and Digital Continuous Education*: The potential design, value proposition, and need for external support regarding innovative and digital continuous education formats to address the various design options available for digital continuous education offerings [2,3,16,49].
4. *Potential Personalization of Continuous Education*: The contextual fit and target group of continuous education formats to reflect on the importance of a domain-specific or target group-specific offering.

Table 1. Guideline Used for the Semi-structured Interviews

No.	Topic	Representative Question
Personal Background and Context		
T1	Personal Information	"Where do you work and what is your role?"
T2	Previous Interactions with Continuous Education Offerings	"With what kind of continuous education offerings have you had experience with so far?"
T3	Responsibility for Continuous Education	"In your current or previous job, do you have responsibility for the continuous education of other employees?"
Status Quo in the Organization		
T4	Current Organizational Procedure	"How is continuous education currently managed in your organization?"
T5	Current Organizational Offering	"What continuous education offerings exist at the moment?"
T6	Affected Personnel	"Which employees should receive continuous education in the area of e-competences?"
T7	Weaknesses of the Status Quo	"Are there any weaknesses in the current continuous education offerings?"
Innovative and Digital Continuous Education		
T8	Offering Design	"How do you think a good continuous education offering, whether digital or analog, should be designed?"
T9	External Support	"Is external support important for an offering?"
T10	Value Proposition and Result	"What is important as a result of continuous education?"
Potential Personalization of Continuous Education		
T11	Contextual Fit	"How important is the specific public sector context in continuous education offerings?"
T12	Target Group	"For which groups of people would you find each of the discussed continuing education formats useful?"
Closing		
T13	Open Issues & Discussion	"Is there anything we haven't discussed so far that is important to you?"

The structure of the used interview guideline with its topics as well as representative questions (translated from German), can be found in Table 1.

To gather our sample, we asked employees from different public sector organizations in Germany with previous experience as participants in continuous education formats that we came across in previous projects to take part in the

Table 2. Roles and Employing Organizations of the Interviewees

Abbr.	Role(s)	Organization
A	CTO & CIO	Statutory accident insurance
B	Head of IT-security & e-government	Municipal administration
C	Advisor E-Government and IT Strategy	State Ministry of Finance
D/E	Organizational development (digitalization)	Municipal administration
F	Project-, Process- & Digitalization Manager	Municipal administration
G	Head of Human Resource Development	Regional corporation under public law

interviews. Some of them work in positions responsible for the education or continuous education of other employees as well (Interviewees A,C,G). Overall, we have conducted six interviews with seven individuals from different governmental organizations. The interviews were conducted and recorded between November 2022 and March 2023 and resulted in 320 min of recording, ranging between 41 and 65 min. Details on the positions and organizations for each interviewee can be found in Table 2.

After these interviews, we stopped the sampling process due to content saturation, i.e., the repetition of aspects regarding our four interview topics. The interviews were conducted and recorded via a remote video call. To evaluate the recordings properly, we transcribed them using a professional online transcription service[2] with some corrective adjustments by the researchers afterward to ensure the correct representation of the interviewees' statements. The transcripts were analyzed using inductive qualitative content analysis, according to Mayring [32], to find similarities and differences for the same aspects in all interviews. Building upon this analysis, we present our general findings regarding the previous experience with continuous education formats and barriers to their utilization in the following Sect. 4.1.

Afterward, we developed a morphological box that defines the properties of continuous education offers for public officials in terms of their content and format. The concept of a morphological box was introduced by the Swiss astronomer Fritz Zwicky [55], and soon gained popularity in other domains, such as information systems [9]. It aims at decomposing complex structures into controllable patterns [42], which is therefore useful to structure the complex properties of continuous education formats in their dimensions in respective characteristics. The discovered dimensions and characteristics of the developed morphological box and their link to the interviewees are described in Sect. 4.2.

4 Results

4.1 Continuous Education and Participation Barriers

The interview participants experienced various continuous education offers before: online-only, presence-teaching, hybrid formats, synchronous online inter-

---

[2] https://sonix.ai.

action with a teacher, platform-based self-study formats, and more. When asked for their individual preference for a specific format, they all agreed upon the use-case-specificity and, therefore, the rationale for the existence of each of these formats. They would, for example, argue that in-person off-site training events are very relevant for acquiring the foundations of a topic and networking with peers, while asynchronous self-study formats are suitable for pure knowledge transfer. However, all participants agreed that the rise of innovative training formats, such as MOOCs, video courses, blended learning, or bite-sized lessons, opens up a whole range of new possibilities, which can help mitigate challenges of current e-competence acquisition undertakings and should be used more intensively.

The participants highlighted a multitude of obstacles to partaking in continuous education. Therefore, this section concentrates on the most severe issues.

The main repeatedly named obstacle is the lack of structure across continuous education formats: due to formats not being attributed with conveyed competences or addressed target groups, it turns out difficult for public officials to choose suitable formats for their individual needs.

"What exists in the market, and is not absolutely obsolete, is completely fragmented and frayed." – Int. A

"They teach Excel and call it public sector digitalization. [...] However, there is nothing that maps competence requirements to public sector digitalization." – Int. A

This also becomes evident in diverging perceptions regarding the overall availability of continuous education opportunities. While one participant argues that there are not enough offers on the market, the others rather highlight the lack of transparency caused by the confusing variety of existing offers.

"And this flood of information is simply insane. In day-to-day business [...] you don't have the time to take care of these things." – Int. B

Moreover, the interviewees noted the often rigid structure of education formats. Considering time-, personnel-, and budget constraints, it is often not possible to partake in off-site training sessions that last longer than one day and require travelling.

"We need low-threshold offerings. If I have to travel to [city] for a week to attend a course like that, then it won't work out. Then it's already difficult. [...] But if I can maybe participate in an event via [conferencing tool] for half a day from home or the office, they think that's better." – Int. B

"One hurdle is certainly, I have to say, the current workload." – Int. G

Therefore, the participants vouched for more formats that use innovative teaching formats and limit off-site training only to cases that make sense in the format.

4.2   Morphological Box as a Means for Structure

Construction: We concluded the analysis as depicted in Sect. 3 with nine distinct dimensions, each with two to four specifications. We additionally enriched the morphological box with one dimension: e-competence classification. The interview participants expressed their desire for attributing the to-be-acquired competences to a continuous education offer, but they were not able to further specify this desire. Hence, we draw on a classification framework for e-competences [22] to depict this requirement.

> "[The expected result from continuous education is] clearly competence building." – Int. C

Subsequently, these ten dimensions were further separated into two supercategories because five of them (Leadership Orientation, Target Group, Entry-Level, E-Competence Classification, Transfer Focus) relate to the content-related orientation, and the remaining five dimensions (Attendance, External Support, Synchronicity, Networking Opportunities, Physical Outcome) relate to the format-related alignment of a given continuous education offer. The distinction between content and format allows for appropriate differentiation when assessing continuous education formats, e.g., for the appropriateness for certain potential learners.

Content: The supercategory *Content* essentially describes *what* a continuous education offer conveys structurally and *for whom* this offer can add value.

Its first dimension, *Leadership Orientation*, points out whether an offer is appropriate for management personnel. The interviewees reported that management buy-in is crucial to implement e-competences on lower hierarchy levels. Leaders must have a certain idea of a topic's importance to facilitate broad deployment.

> "Everyone who is a supervisor, because that is completely lacking in the public sector at the moment. In the public sector, supervisors often do not receive mandatory training in the digital field. A huge problem!" – Int. A

The second dimension, *Target Group*, includes the relevant employee groups that educational offers should differentiate between. Even though digitalization is commonly recognized as a cross-cutting task, the interview participants differentiated between three employee groups, which require varying depths of e-competence in certain fields: the organizational department responsible for, e.g., the organizational architecture and processes, the IT department, responsible for the system architecture and applications in use, and the remaining workforce, handling a variety of (cross-)functional jobs within the operational departments.

> "Then the people in the organization department have a prominent role because they have to carry the message to the country." – Int. B

> "Also from the IT department, because they now realize where the problems of colleagues lie and where the misunderstandings come from." – Int. D

> "When it comes to using certain IT tools, I would say that this now really affects every employee [...] and having the basics of how I really work well digitally and also work safely [is important for everyone]." – Int. G

The third dimension, *Entry Level*, relates to picking up public officials regarding their competence maturity. Based on previous experience and education, public officials might require different aspects of the same e-competence.

The fourth dimension relates to the *E-Competence Classification* is based on Hunnius et al. [22] and was added post the analysis. It aims at creating transparency regarding the conveyed e-competence of an offer.

The fifth dimension, *Transfer Focus*, relates to the previously discussed complexity of (e-)competence acquisition, characterized as a parallel process of providing knowledge and achieving progress in ability. Based on this complexity, the interviewees demanded for continuous education offers to be transparent in whether they focus on conveying knowledge or follow a more practical, application-oriented approach.

> "I think [specific formats] are useful, really, when it comes to such basic things, i.e., to build knowledge foundations." – Int. C

> "I find it very important to bring in case studies, practical cases, so that I can make the transfer better for myself." – Int. G

Format: The supercategory *Format* relates to *how* the content is conveyed. These five dimensions address the personal preferences and potential organizational, professional, and personal circumstances an individual learner is restricted to.

In terms of *Attendance*, the participants indicated a justification for presence-only and digital-only formats, as well as hybrid settings depending on the conveyed content and target audience.

> "Let's put it this way, when you were in presence, the great added value was often the discussions during the breaks." – Int. F

> "What I actually also find really cool are these recordings. I mean, [...] where the lecturer talks about what's going on [and] where you can just watch the videos." – Int. D

> "If I first get an impulse for a topic, then I think digital is simply good, because I can then do it quickly from home [...]." – Int. G

> "I hope that we will be able to combine digital components with face-to-face training [and] that we will be able to better ensure transfer in the future through blended learning measures." – Int. F

The two following dimensions, *External Support* and *Synchronicity*, are closely connected. External Support can be provided, for example, by a human lecturer. Certain digital formats will not provide external support, e.g., when learners consume video-only content. At the same time, this external support does not have to be provided in real-time; monitored forums on MOOC platforms would be an option to provide asynchronous external support.

"It just depends on the target group [and] I would always recommend that [in the case of process management] you do something like what we have now done with this [specific hybrid format]." – Int. A

"In some circumstances, the lecturers, even if they say [they] are available via email or phone,[...] they were quite badly reachable afterwards. [...] With [specific digital format] I can go to the appropriate module and return to it all the time and look at it again." – Int. F

A so far only implicitly addressed important aspect of continuous education for all of our interviewees is the possibility for *Networking Opportunities*. While many argued that this is one of the main benefits of presence formats, one interviewee also argued for the possibility of properly facilitating networking digitally. Logically, e.g., when partaking in unaccompanied and asynchronous digital formats, networking opportunities are not always provided for.

"And I think it's good when I meet people from other administrations [...] and you can exchange ideas. They all think differently and have other solutions at the ready. It's always a good exchange." – Int. B

" In the HR department, I would say that it is also important to exchange information and network with other municipalities. [...] But in the IT area, for example, digital training would make more sense." – Int. E

"Do you know [specific format]? It's also purely digital and it's actually a lot about networking, and they've built it up so cool. [...] Even though it was only online, there was a very strong bond between the group participants." – Int. D

The last dimension depicts the *Physical Outcome* of a continuous education offer. Some participants argued for a certificate of attendance as a minimum outcome of a training format. One participant argued for the importance of formalized acceptance signals, such as qualified performance certificates or even official credit points, such as standardized in the European Credit Transfer System.

"That's not just motivation, but that has a whole different meaning, and a university credit is maximally sexy from my point of view." – Int. C

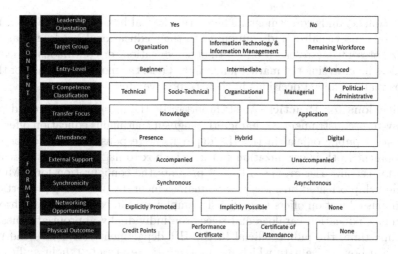

Fig. 1. Morphological Box for Structuring Continuous Education Offers

5   Discussion

The results offer implications for both research and practice. The value of the morphological box, as displayed in Fig. 1, lies within the box's nature to serve as a structuring instrument. As we know from the interviews, a significant drawback of the current landscape of continuous education offers lies within its lack of structure and transparency. Even though appropriate offers might exist, public officials often cannot find them in the overwhelming entirety of offers. Hence, the morphological box is an easy-to-use means to communicate the conceptual frame of continuous education offers transparently.

Implications for Research: The results contribute to the literature stream of e-competence and, thus, to public sector digitalization in the broader sense. As introduced in Sect. 2, scholars are already aware of the importance of public officials for the digital transformation of the public sector [17]. However, current studies examining and conceptualizing e-competence [22,39] have not yet regarded the heterogeneity in the often simplified group of *the public officials*.

Among the EU Member States, approximately 16% and in some countries, even more than 25% of total employees work in the public sector [15], which makes the public sector one of the largest employers with a great variety of employees. Moreover, while some scholars attempted conceptualizing target groups [40], their results did not lead to any real impact. Hence, our results contribute to research because they can help to understand public officials' varying perceptions and demands. The approach of schematically displaying the public officials' perception of (in our case) continuous education offers differs from works like Ogonek et al. [40] as we aim not at grouping the fluid diversity of the workforce in distinct role concepts. Instead, we aggregate viewpoints and perceptions of individuals into a model, which allows for flexible configuration depending on the context and requirements. The morphological box can, for example, help public officials visualize and communicate their educational needs, researchers analyze present-day research on continuous education, or continuous education providers in analyzing the market landscape. Thus, the model helps each of these groups make qualified decisions in their respective area.

Implications for Practice: Most importantly, the morphological box depicts an easy-to-use instrument to characterize a given continuous education offer. For example, a public official in the HR department responsible for e-competence development in their organization can use the box to flag continuous education offers for certain user groups. Figure 2 presents the configuration for a MOOC for leadership-oriented business process management; by presenting properties of continuous education offers so obviously, employees can easily navigate through continuous education catalogs and select the individually appropriate ones. At the same time, the public official in the HR department can utilize the morphological box to evaluate which e-competences are already taught sufficiently and where it is necessary to expand their training landscape. From an education program provider's point of view, the morphological box is an instrument to structure the different offers in a portfolio and thus make it more accessible for potential target groups.

Limitations: Despite the continuous effort put into the research project, some limitations must be highlighted. First, with seven interview participants, our data sample is not necessarily exhaustive. Even though the interviewees agreed upon several aspects and no new structural aspects came up during the last interviews, more research about the generalizability of the findings and suggested morphological box should be conducted. Moreover, our current perspective is only focused on the German public sector and we cannot yet guarantee the applicability of our results in different national contexts with different spe-

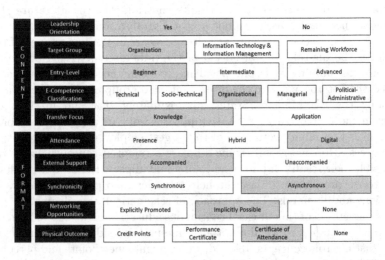

Fig. 2. Morphological Box as Applied to a Process Management MOOC

cialties regarding the structuring of the public sector and its institutions itself. What is more, only by collecting data in the public sector can the findings not be guaranteed to be clearly distinguished from possible results achieved in the private sector. Furthermore, the morphological box is an abstract structuring tool and does not focus on the specific content of continuous education offers. Additionally, as the evaluation of the results is still pending, the morphological box's usefulness, completeness, and comprehensibility still need to be proven, and it should be critically reflected from both academic and practical perspectives.

6 Conclusion and Outlook

Overall, we developed a morphological box that depicts the dimensions and characteristics of continuous education offers focused on e-competences for public officials. The discovered dimensions cover aspects of the offered Content (i.e., Leadership Orientation, Target Group, Entry-Level, E-Competence Classification, and Transfer Focus) as well as the used Format (i.e., Attendance, External Support, Synchronicity, Networking Opportunities, and Physical Outcome). They are based on interviews with public officials utilizing continuous education offers or being responsible for them within their organization.

Future research may evaluate our findings in multiple international settings to look for their general applicability. In addition, other aspects of the offers, the learning- and (e-)competence development process, and necessary e-competences should be researched further to generate impact in practice. Furthermore, developing a solution for structuring existing continuous education offers based on our approach and presenting them to potential learners in a more targeted manner might be helpful for practitioners. Such a solution could also be used to further generate insights into the continuous education landscape by researchers, which can then find gaps and problems with existing offers.

Acknowledgements. This research article has received funding from the projects DFG FOR3539 (GZ: BE 1422/28-1 & GZ: PA 1771/3-1) and the German IT Planning Council / FITKO eGov-Campus 2023 (FI-50/043/001-012023).

References

1. Aikins, S.K., Krane, D.: Are public officials obstacles to citizen-centered e-government? An examination of municipal administrators' motivations and actions. State Local Gov. Rev. **42**(2), 87–103 (2010)
2. Bailey, C., Zalfan, T., Davis, H.C., Fill, K., Conole, G.: Panning for gold: designing pedagogically-inspired learning nuggets. Source J. Educ. Technol. Soc. **9**(1), 113–122 (2006)
3. Bothe, M., Renz, J., Rohloff, T., Meinel, C.: From MOOCs to Micro Learning Activities. In: Proceedings 2019 IEEE Global Engineering Education Conference, pp. 280–288 (2019)
4. Brinkmann, S.: Qualitative Interviewing. Oxford University Press, Oxford (2013)

5. Brunello, G., Wruuck, P.: Skill shortages and skill mismatch in Europe: A review of the literature. EIB Work, Pap (2019)
6. Cerasoli, C.P., Alliger, G.M., Donsbach, J.S., Mathieu, J.E., Tannenbaum, S.I., Orvis, K.A.: Antecedents and outcomes of informal learning behaviors: a meta-analysis. J. Bus. Psychol. **33**(2), 203–230 (2018)
7. Chen, J., Fang, B., Zhang, H., Xue, X.: A systematic review for MOOC dropout prediction from the perspective of machine learning, pp. 1–14 (2022)
8. Chou, T.C., Chen, J.R., Pu, C.K.: Exploring the collective actions of public servants in e-government development. Decis. Support Syst. **45**(2), 251–265 (2008)
9. Couger, J.D., Higgins, L.F., McIntyre, S.C.: (Un)Structured Creativity in Information Systems Organizations. MIS Quart. **17**, 375–397 (1993)
10. De Vos, A., De Hauw, S., Van der Heijden, B.I.: Competency development and career success: the mediating role of employability. J. Vocat. Behav. **79**(2), 438–447 (2011)
11. Dewah, P., Mutula, S.M.: Knowledge retention strategies in public sector organizations: current status in sub-Saharan Africa. Inf. Dev. **32**(3), 362–376 (2016)
12. Distel, B., Ogonek, N., Becker, J.: eGovernment competences revisited - a literature review on necessary competences in a digitalized public sector. In: Proceedings 14th International Conference Wirtschaftsinformatik, pp. 286–300. Siegen (2019)
13. Egloffstein, M.: Massive open online courses in digital workplace learning. In: Ifenthaler, D. (ed.) Digital Workplace Learning, pp. 149–166. Springer, Cham (2018). https://doi.org/10.1007/978-3-319-46215-8_9
14. Estevez, E., Janowski, T.: Landscaping government chief information officer education. In: 2013 46th Hawaii International Conference on System Sciences, pp. 1684–1693. IEEE (2013)
15. Eurostat: share of government employment nearly stable (2020). https://ec.europa.eu/eurostat/cache/digpub/european_economy/bloc-4d.html?lang=en. Accessed 31 Mar 2023
16. Fischer, T., Oppl, S., Stabauer, M.: Micro-credential development: tools, methods and concepts supporting the European approach. In: Wirtschaftsinformatik 2022 Proceedings 1 (2022)
17. Gil-García, J.R., Pardo, T.A.: E-government success factors: mapping practical tools to theoretical foundations. Gov. Inf. Q. **22**(2), 187–216 (2005)
18. Halsbenning, S., Koddebusch, M., Niemann, M., Becker, J.: How to foster e-competence in the public sector? A mixed-method study using the case of BPM. In: EGOV-CeDEM-ePart, pp. 141–151 (2021)
19. Halsbenning, S., Niemann, M., Distel, B., Becker, J.: Playing (government) seriously: design principles for e-government simulation game platforms. In: Ahlemann, F., Schütte, R., Stieglitz, S. (eds.) WI 2021. LNISO, vol. 48, pp. 73–90. Springer, Cham (2021). https://doi.org/10.1007/978-3-030-86800-0_6
20. Hemker, T., Müller-Török, R., Prosser, A.: Interactive eLearning with ERP Systems Advancing Some Refutable Hypotheses on Interactivity in eLearning. In: Central and Eastern European eDem and eGov Days (CEEeGov), Sept. 22, 23, 2022, Budapest, Hungary 1 (2022)
21. Hunnius, S., Schuppan, T.: Competency Requirements for Transformational E-Government. In: 2013 46th Hawaii International Conference System Sciences, pp. 1664–1673 (2013)
22. Hunnius, S., Paulowitsch, B., Schuppan, T.: Does E-Government education meet competency requirements? An analysis of the German university system from international perspective. In: 2015 48th Hawaii International Conference System Sciences, pp. 2116–2123. IEEE (2015)

23. Janowski, T., Estevez, E., Ojo, A.: Conceptualizing electronic governance education. In: 2012 45th Hawaii international conference on system sciences, pp. 2269–2278. IEEE (2012)

24. Koddebusch, M., Halsbenning, S., Kruse, P., Räckers, M., Becker, J.: The increasing e-competence gap: developments over the past five years in the German public sector. In: Fui-Hoon Nah, F., Siau, K. (eds) HCI in Business, Government and Organizations. HCII 2022. Lecture Notes in Computer Science, vol. 13327, pp. 73–86. Springer, Cham (2022). https://doi.org/10.1007/978-3-031-05544-7_6

25. Koddebusch, M., Halsbenning, S., Laude, L., Voss, V., Becker, J.: A song of digitization and law: design requirements for a digitization check of the legislative process. In: Electronic Participation 14th IFIP WG 8.5 International Conference ePart 2022, Linköping, Sweden, Sept. 6–8, 2022, Proceedings, pp. 154–170. Springer (2023)

26. Kresl, P.K.: An aging population and the economic vitality of Pennsylvania's cities and towns (2010)

27. Layne, K., Lee, J.: Developing fully functional E-government: a four stage model. Gov. Inf. Q. **18**(2), 122–136 (2001)

28. Lersch, R.: Unterricht und Kompetenzerwerb. In 30 Schritten von der Theorie zur Praxis kompetenzfördernden Unterrichts. Die Deutsche Schule **99**(4), 434–446 (2007)

29. Lindgren, I., Madsen, C.Ø., Hofmann, S., Melin, U.: Close encounters of the digital kind: a research agenda for the digitalization of public services. Gov. Inf. Q. **36**(3), 427–436 (2019)

30. Littenberg-Tobias, J., Reich, J.: Evaluating access, quality, and equity in online learning: a case study of a MOOC-based blended professional degree program. Internet High. Educ. **47**, 100759 (2020)

31. Margaryan, A., Bianco, M., Littlejohn, A.: Instructional quality of massive open online courses (MOOCs). Comput. Educ. **80**, 77–83 (2015)

32. Mayring, P.: Qualitative content analysis: theoretical foundation, basic procedures and software solution, p. 143 (2014)

33. Mergel, I.: Kompetenzen für die digitale Transformation der Verwaltung. Innov. Verwaltung **42**(4), 34–36 (2020)

34. Mergel, I., Edelmann, N., Haug, N.: Defining digital transformation: results from expert interviews. Gov. Inf. Q. **36**(4), 101385 (2019)

35. Nordhaug, O.: Human Capital in Organizations: Competence, Training, and Learning. Scandinavian University Press, Oslo (1993)

36. Norris, D.F., Moon, M.J.: Advancing E-government at the grassroots: tortoise or hare? Public Adm. Rev. **65**(1), 64–75 (2005)

37. OECD: Recognition of non-formal and informal learning - home. https://www.oecd.org/education/skills-beyond-school/recognitionofnon-formalandinformallearning-home.htm. Accessed 31 Mar 2023

38. Ogonek, N.: The tale of e-government: a review of the stories that have been told so far and what is yet to come. In: Proceedings 50th Hawaii International Conference System Sciences, pp. 2468–2477. IEEE, Manoa, Hawaii (2017)

39. Ogonek, N., Hofmann, S.: Governments' need for digitization skills: understanding and shaping vocational training in the public sector. Int. J. Public Adm. Digit. Age **5**(4), 61–75 (2018)

40. Ogonek, N., Räckers, M., Becker, J.: How to master the "E": tools for competence identification, provision and preservation in a digitalized public sector. In: Proceedings 12th International Conference Theory Practice Electronic Governance, pp. 56–64 (2019)

41. Rhodes, R.A.W.: Recovering the craft of public administration. Public Adm. Rev. **76**(4), 638–647 (2016)
42. Ritchey, T.: Wicked problems-social messes: Decision support modelling with morphological analysis, vol. 17. Springer, Heidelberg (2011). https://doi.org/10.1007/978-3-642-19653-9
43. Rosemann, M., Becker, J., Chasin, F.: City 5.0. Bus. Inf. Syst. Eng. **63**(1), 71–77 (2021)
44. Rowley, J.: Conducting research interviews. Manag. Res. Rev. **35**(3/4), 260–271 (2012)
45. Rusaw, A.C., Fisher, V.D.: Promoting training and professional development in government: the origins and early contributions of SPOD. Public Adm. Quart. **41**, 216–232 (2017)
46. Sarantis, D., Dhaou, S.B., Alexopoulos, C., Ronzhyn, A., Mureddu, F.: Digital governance education: survey of the programs and curricula. Public Adm. Inf. Technol. **38**, 101–119 (2022)
47. Schenk, B., Dolata, M.: Facilitating digital transformation through education: a case study in the public administration. In: Proceedings 53rd Hawaii International Conference System Sciences (2020)
48. Smith, D.A.: An overview of training in the public sector. Working Perspective Education Training, pp. 151–170 (1981)
49. Spector, J.M.: Remarks on MOOCS and Mini-MOOCS. Education Tech. Research Dev. **62**(3), 385–392 (2014). https://doi.org/10.1007/s11423-014-9339-4
50. Stern, S., et al.: Die Besten, bitte: Wie der öffentliche Sektor als Arbeitgeber punkten kann. Tech. rep., McKinsey & Company (2019)
51. United Nations: E-Government Survey 2022: The Future of Digital Government (2022)
52. Weinert, F.E.: Vergleichende Leistungsmessung in Schulen - eine umstrittene Selbstverständlichkeit. In: Leistungsmessungen in Schulen, pp. 17–32. Beltz (2002)
53. Zhang, J., West, R.E.: Designing microlearning instruction for professional development through a competency based approach. TechTrends **64**(2), 310–318 (2020)
54. Zhu, M., Sari, A.R., Lee, M.M.: A comprehensive systematic review of MOOC research: research techniques, topics, and trends from 2009 to 2019. Educational Technology Research Development, pp. 1–26 (2020)
55. Zwicky, F.: Discovery, Invention. Research Through the Morphological Approach, Macmillan (1969)

Institutional Re-design for a Digital Era - Learning from Cases of Automation

Elin Wihlborg[1]([⊠]) [iD], Ida Lindgren[1] [iD], Karin Hedström[2] [iD], and Katarina Gidlund[3] [iD]

[1] Department of Management and Engineering, Linköping University, 581 83 Linköping,
Sweden
elin.wihlborg@liu.se
[2] School of Business, Örebro University, 701 82 Örebro, Sweden
[3] Department of Communication, Quality Management and Information Systems,
Mid Sweden University, 851 70 Sundsvall, Sweden

Abstract. Digital government often addresses how, where, and by whom digital transformation is brought into the complex institutional framed practices of governments and governance of society. However, digital transformation also points at critical demands to address the basic underlying institutional design of governments. The ongoing digital transformation of public welfare institutions opens for a gradual redesign of public institutions. It is important to address core values, such as inclusion, diversity, and literacy, to ensure a reflected transformation and re-design of public institutions. The purpose of this paper is to show how new forms of digital public services may have to be matched with institutional re-design to sustain public values and legitimate governments. We build on Ostrom's eight design principles for institutional governance of common-pool resources and propose four principles for analyzing potential needs for re-design of institutions. Through a re-analysis of two case studies on automation in Swedish public organizations, we illustrate and discuss the institutional design. Hereby, we identify critical points for further analysis of emerging demands for institutional re-design. The analysis indicates that we must see beyond the organizational changes of digital government reforms and programs. We need to stretch into the institutional and foundational models of the public sector and how to provide equal and resilient welfare in a digital and changing world. We conclude by suggesting an agenda for research on institutional re-design in the digital era.

Keywords: Digital government · institutional design · welfare state

1 Introduction

Sustainable public values are grounded in the ideas of inclusive and open societies [26], and guides how to build legitimate institutions [22, 25, 28]. However, digitalization initiatives in public organizations often means that suppliers from tech business bring in market-oriented ways of working in these organizations. As a result, public welfare institutions are gradually being transformed – both consciously and unconsciously – by

N. Edelmann et al. (Eds.): ePart 2023, LNCS 14153, pp. 99–113, 2023.
https://doi.org/10.1007/978-3-031-41617-0_7

new technologies, networks, data sharing across organizational and national borders, and new ways of performing services and work [5]. It has been argued that this so-called digital transformation gradually also transforms how our contemporary society is organized, [8], which gives rise to a need to take this further and open for a redesign also of the institutions in society.

The ongoing digital transformation of public welfare institutions entails a gradual re-design of social and political institutions, and there is a need to pinpoint public values [6, 11, 24, 29] to avoid a transformation of our public institutions in an uncritical way. If such public values are not acknowledged, friction can arise when tactics and strategies designed for industry and commercial organizations are transferred into public sector [21], and the design of welfare and social services risk adopting profit-maximizing business models unreflectively. This, in turn, risks undermining the foundation of the welfare models that has promoted economic as well as social sustainability. There is a need to open up and challenge the institutional framing of welfare to see how digital transformation fundamentally can change processes and delivery of welfare. We will look beyond single cases to discuss general implications for institutional design beyond the contemporary boxed institutional arrangements [2].

To exercise rights and citizenship and participate fully in society today, requires that people can access and handle various digital tools and services. Also, digitalization has changed the time and place of public encounters between government and citizens [25]. For most citizens, it increases flexibility and accessibility, but new societal divides have simultaneously and unintentionally been created between those included, and those excluded, by digital public services, seen as a digital divide [19]. These challenges are even more explicit in advanced welfare societies as the Scandinavian welfare states. Based on values such as equality, impartiality, and legitimacy [26, 28, 30], Scandinavian welfare organizations provide extensive services to citizens, from the cradle to the grave. Guided by directives to provide access to service through digital channels to the extent possible, digital self-service solutions is the preferred mode of communication with citizens. This has transferred tasks previously conducted by welfare organization professionals to the citizens [17], reinforcing the digital divide. In addition, the increasing use of automation in the administration of public service [1, 13] challenges the institutional framing and its guidance for organizing daily welfare services.

There is a need to leverage the Scandinavia welfare model and its institutional design in line with the digital transformation, without compromising its core values. Our welfare models initially became organized around participation in and contributions to the labor force, as a base for the extensive social insurance benefits, public services, and the large magnitude of income redistribution [9]. This welfare model was designed as a way to tame the industrial model of technology, by working time arrangements, high degree of decommodification of work, and educational opportunities. The industrial era organization of work was regulated in a particular time and space, and this framed the models for the welfare institutions. Today, the so called 'liquid modernity' [3] or 'second modernity' [4] of the digital era deeply embeds digitalization (the technology of our time) in the public organizations and peoples' lives, meaning that digital technologies have dissolved previous time-space related boundaries between organizations, and between work and spare time. Contemporary reframing of work, living conditions, and public arrangements face new challenges demanding new institutional settings. Here, the Scandinavian welfare states is a relevant context to look for new institutional re-design of welfare in the digital era.

The purpose of this paper is to unveil how digitalization re-arranges institutional arrangements, improving our understanding of sustainable institutional design for welfare in the digital era. We use automation in public services to illustrate our overall argument that there is a need for a critical approach to public sector digitalization to secure, sustain, and develop how public values are guiding institutional design in a digital era. To do so, we reanalyze and synthesize two empirical case studies on automation as a critical illustration of digital transformation in public welfare systems. The analysis is guided by Ostrom's framework on sustainable institutional design [22]. In contrast to Ostrom's design principles, where the common resources lack institutional framing, the digital transformation challenge established institutional frames of welfare. Thus, we will here elaborate on a reversed model of institutional design principles, showing how to form new institutional arrangements around a new common resource. We thus open for a discussion on institutional re-design with a clear starting point in welfare models and institutional practices.

This paper proceeds with three main parts. Firstly, we ground the argumentation in a theoretical framework on institutional design and our methods used for reanalysis and synthetizing the cases. The second part of the paper presents the two strategically selected cases that are reanalyzed to highlight their implications for institutional design. The third and final part of the paper compares the implications from the case studies and open for implications on how to elaborate on a research agenda for sustainable institutional design in the digital era.

2   Institutional Design for a Digital Era - Theoretical Foundation

To discuss how digitalization re-arranges institutional arrangements, scope our revisited cases studies, and identify underlying institutional challenges and components, we use Ostrom's framework on sustainable institutional design [22] that is commonly used also in relation to public administration analysis [33].

2.1   Institutional Theory and the Roles of Government

In social science, institutions are typically presented as social structures or systems that are created by human beings to support purposes or functions in society. Institutions are decided upon by formal decision-making agencies or through informal norms and practices [29]. Institutions, as abstract entities supporting and guiding how to live together in societies and how to make the most out of it, are reducing the risks of living together and make interactions smoother by reducing transaction costs [20]. In practice, institutions can take many forms, including government organizations, educational systems, religious institutions, financial institutions, and legal systems. Institutions evolve over time, and shape individual and group behavior in relation to changes in societies, such as new technologies [18]. Institutions interact with one another, they are influenced by political and economic forces, and can be reformed or transformed to better serve the needs of society.

Overall, institutions are seen as a fundamental aspect of social organization and governance, and social scientists seek to understand their role in shaping society and the behavior of individuals within it. Institution is a set of established rules and practices that shape social behavior. In social science, the terms "organization" and "institution" are therefore often used interchangeably, even though they actually have different meanings. An organization is more limited and framed within the institutional arrangements in the specific context. Organizations can also be more or less formal and consist of a group of people who work together to achieve a common goal. Both public and private organizations are formed for a specific purpose and are typically created to achieve a specific objective, such as producing goods or providing services.

Since organizations are formed and take place within an institutional setting, institutions work on a larger scale and over longer periods of time. Institutions are more abstract and refer to the formal and informal rules and practices that govern social behavior. They can include formal organizations, but also broader social norms, customs, and traditions that shape behavior. Examples of institutions include the legal system, the educational system, and the family. Institutions can be seen as a way to stabilize social behavior and provide a sense of predictability and order. Since institutions are more enduring and stable than organizations, their design is more complex and complicated and must be more sustainable and trustworthy. Institutions are deeply embedded in social structures and are difficult to change, while organizations are often more adaptable and can change more quickly in response to new circumstances. By focusing on institutional design [12] from a socio-technical [7] framing, we acknowledge the formal, informal, and cognitive functions of institutions.

2.2 Institutional Design

Institutional design aims to re-shape institutions to support aims and values more efficiently, or in the language of economists – to reduce the transaction costs [20]. The institutions around a public educational system, like funding, curriculum, examinations and evaluation, may be adopted to demands from and fluctuations on the labor market. As in times of high employment, most welfare states provide some types of additional funding for vocational training to increase employability. Such institutional design may be temporal, until the economy is more in balance, and it builds on all parts of the institutions around them. When new institutional arrangements are introduced, actors at the micro-level adjust their behaviors and seek to interpretate new generalized meanings of the macro-level arrangements [27 p. 87]. Institutional design provides a new form of institutional configuration that in turn delivers a different array of capacities for those involved with micro level practices framed by the institutions. Local stakeholders, who knows the norms and values of the community, will arrange the institutions based on the characteristics of the resource to meet the need of the community [23 pp. 238–39]. The aim to be resource efficient is not simply an economic calculation but includes community values more broadly.

Institutional change, on a structural level, interplays with local actors in organizations that adopt to and innovate within the context, since actors are embedded in ongoing processual conceptualizations [23]. Geels [10] concludes that institutional arrangements around socio-technical transitions are evolutionary processes given meanings through interpretive and socio-cultural processes. He shows how niches for innovations are opened through the institutional framing, and digital government changes can be seen as innovations in an organizational niche formed by the evolutionary process of how to govern the commons.

2.3 Governing the Commons – Ostrom's Framework for Institutional Design

Institutions form how we govern common resources, as shown by Ostrom's eight design principles that are crucial for effective institutional governance of common-pool resources [22]. These design principles are:

- *Clearly defined boundaries*: Institutions should have clear boundaries that define who is and who is not a part of the resource system being managed.
- *Proportional equivalence between benefits and costs:* The costs of using a resource should be proportional to the benefits derived from that use.
- *Collective choice arrangements*: The individuals who use a resource should be able to participate in the decision-making process regarding its management.
- *Monitoring:* The resource system should be monitored to ensure that it is being used appropriately and that any violations of the institutional rules are detected.
- *Graduated sanctions:* Punishments for violating institutional rules should be graduated according to the severity of the offense.
- *Conflict resolution mechanisms:* Institutions should have mechanisms in place for resolving conflicts between users of the resource.
- *Minimal recognition of rights*: The institutional rules should be designed to recognize the right of individuals to use the resource and to participate in its management.
- *Nested enterprises*: The governance of the resource should be nested within larger governance structures, allowing for the coordination of multiple resource systems and ensuring that the resource is being managed in a way that is consistent with broader social goals.

In contrast to Ostrom's design principles, where the common resources lack institutional framing, the digital transformation challenge established institutional frames of welfare. Thus, we will here elaborate on a reversed model of institutional design principles, showing how to form new institutional arrangements around a new common resource.

3 Analytical Framework to Critically Reflect Upon Institutional Re-design

Ostrom's [22] principles are intended to guide the design of institutions and can effectively manage new common-pool resources over the long term, while promoting cooperation and equitable outcomes for all users of the resource. The model is developed to understand and to promote how institutions are formed in new settings, or when new conflicts around a resource appear. However, we use these to elaborate on the re-design of institutions, since there are institutions formed to arrange welfare that are now challenged of digital transformation as shown through the case of automation below.

We use Ostrom's design principles in the reverse order to identify re-design of existing institutional arrangements. Since digital government innovations are made in organizational niches [10] with high degree of private involvement through available systems, striving for efficacy, there is a need to address the structural interpretations of values in the institutional re-design. The difference between organizational change and institutional re-design, makes us use organizational niche innovations of automation in public services as an illustration of the need for institutional re-design. There is a stability and slowness inherent in the concept of institutional change that needs to be taken into account not only as a hurdle for digital transformation but also as an important foundation to guarantee sustainability and transparency.

The resources at stake in the welfare systems are the time, money, and support provided to those in need. These are taken from the recourses pooled through the taxes paid. However, for the individual, the welfare system is also a resource when not used, as it can be perceived as a resource to know that there is an income maintenance support system, study grants, health care, or pensions available when needed. It reduces the risks of living and bring the community together. The welfare system is a common pool resource that has to be framed in new institutions due to the digital transformation of the access to and use of the common resources. Thus, we reverse the design principles to form a tool for analyzing re-design of institutions. We propose that these can be used to start the analysis in existing structures of institutions to identify what hampers digital transformation:

- **Challenges and obscurities in the nested organizations.** Challenges and obscurities in the nested organizations indicate a need for re-coordination of the multiple resource systems, to find a new arrangement where resources can be managed in a way that is consistent with broader social goals. (Reframed version of point 8.)
- **The recognition of rights and duties are not clear.** The recognition of rights and duties are not clear, and there is a need to search for rules that can recognize the right of individuals to use the resource and to participate in its management. (Reframed version of point 7.)
- **Conflicts of Interest Concerning the Institution.** When conflicts, both open and potential conflicts, are not resolved, the monitoring and sanctions are unclear and lack consensus. (Reframed version of point 6, 5 and 4.)
- **The "resource" – what is framed and used within the institution.** The "resource" is not clear or defined, and the users lack knowledge and competence to understand boundaries, costs, and benefits of the resource. The user can neither visualize nor participate in the decision-making regarding the management. (Reframed version of point 3, 2 and 1.)

4   Empirical Illustrations – Revisiting Two Cases

We now turn to two empirical cases and apply the four re-design principles to these cases.

4.1 Case 1: Automation of Case Handling in a Local Government

The first re-visited case study focus on the implementation of Robotic Process Automation (RPA) for streamlining case handling in a Swedish municipality (i.e., local government). It is part of a larger project that seeks to (1) map current implementations and use of RPA for automated case handling in the municipality, and (2) develop an analytical tool that can be used by researchers and practitioners to decide if, and to what degree, a specific case handling process can (and should) be automated [15].

The Setting of the Case Study

Swedish municipalities are self-governing, with a strong local autonomy for institutional arrangements within the frames of national legislation. Many of these municipalities are currently facing budget cuts and a growing population, creating a need for cost-reductions. RPA is highlighted as a possible means to make local government more efficient and effective, as expressed by both policy makers and IT experts. Consequently, many Swedish municipalities are experimenting with RPA as an administrative tool, to see whether RPA can reduce the need for human labor in the administration of public services [17]. The case concerned a municipality's work on RPA implementation. The case study built on a qualitative and interpretive approach [17, 36]. We conducted 21 interviews with 18 respondents (between February 2020 and January 2021). The respondents were spread across different departments within the municipality and were predominantly working as managers or business developers. Analyses and results from this case are published in [14, 31, 34].

The Case Study Results

The case highlights several challenges associated with institutional design [16]. RPA implementation was initiated in the organization from the top-down without bottom-up support, meaning that external and top-management pressures guided the implementation of RPA in the organization. In the organization, however, there was not sufficient process- and IT-competence to successfully work with RPA. This, in turn, created a dependence on individual enthusiasts and external RPA consultants. Without support from the employees whose work content could potentially be supported by RPA, difficulties arose in finding the appropriate processes to automate. The case illustrates the tension between top management's abstract digitalization visions and the experience-grounded ideas originating from e.g., case handling staff, and the tensions springing from responsibilities and power being distributed across RPA stakeholders [14]. The case further illustrates how the municipal context shapes the development of RPA in the organizations observed. For example, Swedish municipalities are self-governed multi-service providers containing independent authorities, administrations, and heads of administration, as well as professional case handling staff. The distribution of power and responsibilities across and within the organization challenges traditional ways of developing IT and call for alternative organizing principles to realize RPA [14].

Implications for Institutional Design

Using the vocabulary in the re-design principles, we can see RPA as a new resource in the organization; a resource that, in turn, requires other resources (personnel, time, and money) to be realized. As illustrated briefly, the realization of RPA was made difficult by the nested, multi-departmental organization of the municipality and calls for new arrangements on how to finance, plan and coordinate RPA development in the organization. In the current organization of RPA implementation in the municipality at hand, the rights and duties in relation to RPA development were not clear. Also, it was unclear if all necessary stakeholders were able to participate in the management of the pooled resources needed to realize RPA. Analyses showed that the resource – RPA, and the other resources needed to realize the potential of RPA – were not clearly defined. In fact, different stakeholders in the organization had very different understandings of RPA, where some had very little knowledge of its boundaries, costs and benefits [31]. Conflicts between different stakeholders' views were visible and not dealt with [32].

4.2 Case 2: Challenging Accountability When RPA is Introduced

The second re-visited case study focused on the introduction of automated decision making, a form of RPA, in a national agency in Sweden and we focused on how accountability was reframed and demanded new institutional arrangements.

The Setting of the Case Study

The case study was conducted at the Swedish Transport Agency (STA), who is responsible, among several other issues, for issuing driver licenses in Sweden and keeping the Swedish vehicle record. STA is one of the most digitalized Swedish agencies. They have not only, as most other agencies, digitalized and automated large parts of its internal case handling, but also fully automated several public services and decisions, such as decisions about driver license learner's permit. Two of the authors followed the processes on how automatic decision-making changes procedures and practices among the caseworkers, and its implications on relationships with the clients [35]. Through the large number of client contacts that was handled, we could investigate how accountability was constructed in the interface between the government officials and the clients and in addition how the automated system interfered these relations.

The Case Study Results

In the case of revocation of a driver's license, the STA assesses personal circumstances that have importance for the individual's suitability. To assess the severity of a revocation, the STA may also examine other personal circumstances such as if the person needs the car for work, their living situation, etc. The STA has implemented RPA that manages applications about driver supervision, driver license permits, and re-application of driver licenses for heavy vehicles. The automatic system prepares and handles cases, but it also approves or dismisses a license based on the National Driver License Act (SFS 1998:488), Driving ordinance (1998:980), and the STA's regulation about medical requirements for driver license. The automatic system is thus working based on a rule-based implementation, strictly following programmed rules. The locus of responsibility is changing when the public decision-making becomes automated. Automatic decision-making re-arranges relations concerning accountability, changing the balance of power. The locus of responsibility can change in two ways. The main re-location was to the RPA, who became responsible for all standard errands follow-ing normal predicted lines of decision. The RPA was in all these cases both making and communicating the decision. The RPA thus has a delegated responsibility and is assigned accountability. However, when an individual case did not follow the lines of decision inscribed into the RPA, the locus of responsibility was transferred to a human case worker who could gain increased flexibility in the decision based on professional competences and service-oriented values seeing the human client behind each case. The human case workers could speed up errands and include other aspects, such as personal concerns and special cases, into their decision making.

Implications for Institutional Re-design
The organization around RPA in this case demonstrates a great deal of ambiguity. Leg-islation and formal regulations are given new meanings when translated into the RPA system. The prescribed routines and rules can be viewed as a working order, where the computer has been attributed a great deal of agency, executing the underlying rules. We could see how the RPA acted as a delegate on the behalf of the governmental deci-sion maker(s) as part of "organizing accountabilities" [35]. This is in line with what Woolgar and Neyland [35 p. 39] identified also in systems with less, or even no, inter-nal capacity to act, as traffic lights in their cases. in this case the power delegated into the RPA re-arranges the power relations among the involved actors and the formats for accountability, beyond the ordinary institutional arrangements.

These RPA systems might seem strong and powerful, but from a socio-technical view, the systems are re-framing relations in the network where they are residing, that in turn has implications for accountability. Computers in general, and robots in particular, are not only supporting the automation of administrative processes, but are also attributed agency in their role as decision-makers [14, 35]. They are therefore included in a complex chain in relation to accountability that was re-arranged. Such a system is interpreted and become part of a sense-making accountability practice, having a powerful effect and ruling, not only on the working-life of the public sector officials, but also the lives of citizens. This forms a more complex arrangement where the contextual setting of the RPA extends how we interpretate accountability, beyond the simple argument that robots cannot be accountable since they lack intentions. The nested organizations are not clearly integrated, on individual issues based on medical information, criminal records etc. that have importance for getting a driver license permit.

5   Discussion on Implications for Institutional Re-design

The purpose of this paper is to unveil how digitalization re-arranges institutional arrange-ments, improving our understanding of sustainable institutional design for welfare in the digital era. Using the examples of automation in public services above, we now proceed to discuss the need to form new institutional arrangements around a new common resource, based on the four proposed re-design principles.

5.1   Challenges and Obscurities in the Nested Organizations

Based on Ostrom [22], we claimed that challenges and obscurities in the nested organi-zations indicate a need for re-coordination of the multiple resource systems, to find a new arrangement where resources can be managed in a way that is consistent with broader social goals. In both cases there were almost paradoxical indications of the outcomes of the processes and a need for re-coordination of the multiple resource systems. These challenges demand new arrangements where resources can be managed in a way that is consistent with broader social goals.

Both cases show that processes of introducing RPA pointed at challenging aspects that were beyond the reach of the professional roles typically involved in digital government initiatives, such as case workers, business developers, and IT personnel. As illustrated in the first case, each municipality interested in using RPA must procure the necessary technology separately. When procured by the municipality, the RPA procured cannot be easily used by all stakeholders in the organization, i.e., the independent authorities, administrations, and heads of administration, making up the municipality as a whole. The involved stakeholders did not know how to coordinate resources across the nested parts of the municipality to realize the potential of RPA. Also, as nested organizations there is a risk for lack of transparency in the integration of different organizational systems. In the case of the national agency STA, this was illustrated by the black boxing of information caused by the RPA as it collected information from different systems in order to take a decision about driver license permit. As a result, the professional case workers experienced that they could not be accountable for their work since the RPA constrained their discretion to adjust decisions. In additions both cases showed that the distribution of power and responsibilities across and within the organization were challenged by the alternative organizing principles to realize RPA.

5.2 The Recognition of Rights and Duties are Not Clear

Based on the principles by Ostrom [22], we claimed that when recognition of rights and duties are not clear, there is a need to search for rules that can recognize the right of individuals to use the resource and to participate in its management. In the cases above we have seen that there are challenges of transparency in the decision-making process, and there is a need to search for rules that can recognize the right of individuals to use the resource and to participate in its management.

In the first case, the rights and duties of various internal stakeholders in the municipality, in relation to RPA development and implementation, were not clearly defined [14]. For example, those stakeholders responsible for driving the development of RPA in the organization did not have financial resources or decision mandate regarding RPA; instead, the resources needed to realize RPA were distributed to other stakeholders with little insights on RPA [14]. Lack of communication and transparency between various parts of the organization made RPA development come to a halt [16].

In the second STA case, we see a different and more hands-on variant of this phenomenon. The RPA was introduced to standardize the handling of single cases to make them more impartial, and the management at STA expected the RPA to be a 'digital co-worker' in the case management process. The human case workers had to make all negative decisions; thus, the collaboration was never able to be between equal actors. The case worker had to support the RPA on tricky issues but did not get the similar support. Furthermore, since the human case workers could add aspects of personal issues and special concerns when making their decisions, they had power to stretch beyond the formalized impartiality coded into the RPA. It indicates that the human case worker and the RPA could recognize the clients' rights and duties differently. Thus, the rights and duties of the various actors in the decision process are not equal and clearly defined.

5.3 Conflicts of Interests Concerning the Institution

Based on the principles by Ostrom [22], we claimed that institutional re-design may be needed when conflicts, both open and potential, are not resolved, and the monitoring and sanctions are unclear and lack consensus. In the first local government case, it was clear that there was a conflict of interest appearing in the interface of whether or not RPA was a valuable new resource for the municipality or not. Without structures for coordinating this type of cross-departmental technology, conflicts of interests arose, making monitoring of the development difficult as well. On a higher abstraction level, we also see that the self-governed municipalities have formed local practices, norms and routines that now became challenged by the standardization that RPA entails. This is seen also in the second case, where the use of RPA made it clear that the national legal norms had to be re-interpreted in order to be programmed into the system. As a result, the RPA and the human case workers resolved their cases differently. Similarly, at the STA disputes appeared along the strict use of the legal arrangements in the RPA on the one hand, and the case workers discretion on the other. The case workers had a clear demand to be able to use their professional competences and make individual adjustments in the decisions. Before the introduction of the RPA, the demands among the case workers had rather been the other way around, striving to find standards.

5.4 The "Resource" – What is Framed and Used Within the Institution

Lastly, we claimed that there may be cause for institutional re-design if the "resource" is not clear or defined, and the users lack knowledge and competence to understand boundaries, costs, and benefits of the resource. The user can neither visualize nor participate in the decision-making regarding the management. In both cases, we see how the involved stakeholders have different views on the nature of RPA. Is RPA a new common resource to be pooled? If so, other resources (personnel, time, and money) are needed for RPA to be realized as a new resource. In the first case, necessary shared structures for pooling the resources needed to realize RPA as a resource in the municipality, were not put in place. As a result, each sub-department of the municipality had to finance its own RPA development, thus missing out on the potential scale-up benefits of developing RPA as a shared resources across departments. Analyses showed that the stakeholders involved lacked the necessary competence to understand the boundaries, costs, and benefits of RPA as a resource.

Ostrom [23 pp. 238–39] claims, as said above, that local stakeholders will arrange the institutions around the resource based on their core values, and in both these cases it was complicated since the partially conflicting public and market based values were not combined. Characteristics of the resource to meet the need of the community. The aim to be resource efficient is not simply an economic calculation but includes community values more broadly. In the second case, it was unclear how the local community (the human case workers; and the programmers of the RPA system) have arranged the institution to meet the welfare aims of the community in the design of the RPA system and the overarching decision-making process. Thus, there is a need to further elaborate on and discuss policy implications on how automatization and digital government in general reframes the institutions of welfare in the digital era. It is obvious that one key resource can be the RPA itself, but currently without clear boundaries for where the RPA starts and ends.

6 Concluding Remarks

The four principles, as we developed based on Ostrom's model for governance of common pool resources, has shown a potential for analyzing the need for institutional re-designed and a potential to identify and highlight institutional incompatibilities. Our main argument, encouraged by Bannister [2] is that we have to look beyond the contemporary institutional framing to see how RPA and digital government more generally can support public values and sustainable inclusive societies.

The four reversed steps for institutional re-designed, as used above were:

- **Challenges and obscurities in the nested organizations,** that was shown through how RPA could not become embedded into neither the national agency nor the self-governed municipality organization. In spite of high ambitions, there were both professional challenges, and governance obscurities in the organizations supposed to use the PRA. Both organizations were struggling with re-coordination of the multiple resource systems, but the governance tools were beyond their reach and those in power here could not find or design new arrangement where resources can be managed in a way that is consistent with broader social goals. There were obvious risks of reducing key public values as legality, justice and accountability when the commercial providers of the RPA got power to design and direct the implementation.
- **The recognition of rights and duties are not clear.** In both the re-analyzed cases, the professionals and the management identified that rights and duties had to be reframed, both for professionals and other users. The role of the RPA was not clear, since it's rights and duties had not been clarified and risked of being more guided by market values as efficiency and productivity than public core values as legality, justice and accountability.
- **Conflicts of Interest Concerning the Institution.** There were conflicts in both the case studies, but these involved different stakeholders and were more open in the municipal case than at the national agency. None of the organizations had any clear models for monitoring use and outcomes of the RAP. In general terms the involved stakeholders lacked consensus on the reasons for introducing RPA and how the work together with it.
- **The "resource" – what is framed and used within the institution.** All the above discussed challenges of the RPA introduction can be reduced to their unclear and undefined resource. The actors in both settings and with different roles in the organizations, in particular policy makers, lacked knowledge and competence to understand boundaries, costs, and benefits of the resource. We could even conclude that they lacked explicit reasons and motives to introduce the RPA and what public values it had a potential to improve.

Through our case studies, as re-analyzed in the institutional design framing above, we can see that there is a need for further emphasis how to bring public values in the forefront of the development of new practices and institutions in digital government. Three key values that we identified as risks above, and that are critical to enhance are legality, justice and accountability. Legality is a stake when the professional competences lose discretion and when impartiality is at risk. Justice is a more general value, that has to be related to clear resource pools to distribute and as long as the resources is not defined it is not possible to discuss and manage a just and fair distribution of the resource. Accountability is problematic as long as stakeholders cannot be held accountable, and sanctions are therefore tricky since they are not involved in the design. How accountability is divided among public actors and commercial providers of the technology is neither addressed in the current institutional setting. Thus, we have to look beyond the boxes formed by our contemporary institutional arrangements for resilient and sustainable digital government. As we have learned from Ostrom, the values and interests embedded in a resource such as an RPA will frame how professionals and management construct and re-arrange their work relations in line with the institution. As digitalization in public organization not seldom borrows ideas from tech business, there are therefore risks that we build institutions on a foundation of market-oriented values.

The conceptual discussions and proposed model in this paper can be used to open for more synthesizing studies grasping theories and implications beyond single case studies. We will show how it is possible to stretch over contextual factors, in novel ways, to elaborate on institutional design framework that has a potential to guide both research and support practical institutional design in real life. Our re-analyses of the cases show that challenges and problems, are often related to institutional design, beyond the organizational setting and frames. We propose to open the analyses beyond the current institutional framing of digital government to build, test and elaborate on new theoretical models on resilient institutional design in the digital era. When analyzing digital government projects, there is also a need to see the structures provided by institutions open for re-design. There is a need for new models, useful for both research and practice, striving to enhance and grasp institutional design more suitable for the digital era.

References

1. Andersson, C., Hallin, A., Ivory, C.: Unpacking the digitalisation of public services: configuring work during automation in local government. Gov. Inf. Quart. **39**(1), 101662, 1–10 (2022). https://doi.org/10.1016/j.giq.2021.101662
2. Bannister, F.: Beyond the box: reflections on the need for more blue sky thinking in research. Gov. Inf. Q. **40**(101831), 1–11 (2023). https://doi.org/10.1016/j.giq.2023.101831
3. Bauman, Z.: Liquid Modernity. Polity Press, Malden (2000)
4. Beck, U., Lau, C.: Second modernity as a research agenda: theoretical and empirical explorations in the 'meta-change' of modern society. Br. J. Sociol. **56**(4), 525–557 (2005). https://doi.org/10.1111/j.1468-4446.2005.00082.x
5. Benner, M.: The Scandinavian challenge: the future of advanced welfare states in the knowledge economy. Acta Sociologica. **46**(2), 132–149 (2003). https://doi.org/10.1177/0001699303046002004

6. Bernhard, I., Gustafsson, M., Hedström, K., Seyferin, J., Whilborg, E.: A digital society for all? - Meanings, practices and policies for digital diversity. In: 52nd Hawaii International Conference on System Sciences (HICSS), pp. 3067–3076 (2019). https://doi.org/10.24251/HICSS.2019.371
7. Bijker, W.E.: Of Bicycles, Bakelites, and Bulbs: Toward a Theory of Sociotechnical Change. MIT press, Cambridge (1997)
8. Collington, R.: Disrupting the welfare state? Digitalisation and the retrenchment of public sector capacity. New Polit. Econ. **27**(2), 312–328 (2022). https://doi.org/10.1080/13563467.2021.1952559
9. Esping-Andersen, G.: The Three Worlds of Welfare Capitalism. Princeton University Press, Princeton (1990)
10. Geels, F.W.: Micro-foundations of the multi-level perspective on socio-technical transitions: Developing a multi-dimensional model of agency through crossovers between social constructivism, evolutionary economics and neo-institutional theory. Technol. Forecast. Soc. Chang. **152**, 119894 (2020). https://doi.org/10.1016/j.techfore.2019.119894
11. Gil-Garcia, J.R., Dawes, S.S., Pardo, T.A.: Digital government and public management research: finding the crossroads. Public Manage. Rev. **20**(5), 633–646 (2018). https://doi-org.e.bibl.liu.se/https://doi.org/10.1080/14719037.2017.1327181
12. Goodin, R.E. (ed.): The Theory of Institutional Design. Cambridge University IPress, Cambridge (1998)
13. Larsson, K.K.: Digitization or equality: when government automation covers some, but not all citizens. Gov. Inf. Quart. **38**(1), 101547, 1–10 (2021). https://doi.org/10.1016/j.giq.2020.101547
14. Lindgren, I., Åkesson, M., Thomsen, M., Toll, D.: Organizing for robotic process automation in local government: observations from two case studies of robotic process automation implementation in Swedish municipalities. In: Juell-Skielse, G., Lindgren, I., Åkesson, M. (eds.) Service Automation in the Public Sector: Concepts, Empirical Examples and Challenges, pp. 189–203. Springer, Cham (2022). https://doi.org/10.1007/978-3-030-92644-1_10
15. Lindgren, I.: Exploring the use of robotic process automation in local government. EGOV-CeDEM-ePart **2020**, 249–258 (2020)
16. Lindgren, I., Johansson, B., Söderström, F., Toll, D.: Why is it difficult to implement robotic process automation? Empirical cases from Swedish municipalities. In: Janssen, et al. (eds.) EGOV 2022. LNCS vol. 13391, pp.353–368. Springer, Cham (2022). https://doi.org/10.1007/978-3-031-15086-9_23
17. Madsen, C.Ø., Lindgren, I., Melin, U.: The accidental caseworker – how digital self-service influences citizens' administrative burden. Govern. Inf. Quart. **39**(1), 101653, 1–11 (2022). https://doi.org/10.1016/j.giq.2021.101653
18. Nelson, R.R.: An Evolutionary Theory of Economic Change. Harvard University Press, Harvard (1985)
19. Norris, P.: Digital Divide: Civic Engagement, Information Poverty, and the Internet Worldwide. Cambridge University Press, Cambridge (2001). https://doi.org/10.1108/146366903322008287
20. North, D.C.: Toward a theory of institutional change. Political Econ. Inst. Compet. Represent. **31**(4), 61–69 (1993)
21. Nyhlén, S., Gidlund, K.L.: In conversation with digitalization: myths, fiction or professional imagining? Inf. Polity. **27**(3), 331–341 (2022). https://doi.org/10.3233/IP-200287
22. Ostrom, E.: Governing the Commons: The Evolution of Institutions for Collective Action. Cambridge University Press, Cambridge (1990)
23. Ostrom, E.: Understanding Institutional Diversity. Princeton University Press, Princeton (2005)

24. Panagiotopoulos, P., Klievink, B., Cordella, A.: Public value creation in digital government. Gov. Inf. Q. **36**(4), 101421 (2019). https://doi.org/10.1016/j.giq.2019.101421
25. Pollitt, C.: New Perspectives on Public Services: Place and Technology. Oxford University Press, Oxford (2012)
26. Popper, K.: The Open Society and its Enemies. Routledge, London (1945)
27. Room, G.: Complexity, Institutions and Public Policy: Agile Decision-Making in a Turbulent World. Edward Elgar Publishing, Cheltenham (2011)
28. Rothstein, B.O., Teorell, J.A.: What is quality of government? A theory of impartial government institutions. Governance **21**(2), 165–190 (2008). https://doi.org/10.1111/j.1468-0491. 2008.00391.x
29. Scott, W.R.: Institutions and Organizations: Ideas, Interests, and Identities. Sage publications, Thousand Oaks (2014)
30. Svallfors, S. (ed.): Contested Welfare States: Welfare Attitudes in Europe and Beyond. Stanford University Press, Stanford (2012). https://doi.org/10.11126/stanford/9780804782524. 001.0001
31. Toll, D.: Sociotechnical imaginaries of the automated municipality. Licentiate dissertation, Linköping University Electronic Press (2022)
32. Toll, D., Lindgren, I., Melin, U.: Stakeholder views of process automation as an enabler of prioritized value ideals in a Swedish municipality. JeDEM **14**(2), 32–56 (2022). https://doi. org/10.29379/jedem.v14i2.726
33. Toonen, T.: Resilience in public administration: The work of Elinor and Vincent Ostrom from a public administration perspective. Public Admin. Rev. **70**(2), 193–202 (2010). https://doi-org.e.bibl.liu.se/https://doi.org/10.1111/j.1540-6210.2010.02147.x
34. Wihlborg, E., Larsson, H., Hedström, K.: "The Computer Says No!" – a case study on automated decision-making in public authorities. In: 49th Hawaii International Conference on System Sciences (HICSS), pp. 2093–2912. IEEE Computer Society (2016). https://doi.org/ 10.1109/HICSS.2016.364
35. Woolgar, S., Neyland, D.: Mundane Governance: Ontology and Accountability. Oxford University Press, Oxford (2013). https://doi.org/10.1093/acprof:oso/9780199584741.001. 0001
36. Zürn, M.: A Theory of Global Governance: Authority, Legitimacy, and Contestation. Oxford University Press, Oxford (2018). https://doi.org/10.1093/oso/9780198819974.001.0001

Digital Technology

How Search Engines See European Women

Kristian Dokic[(✉)] [iD], Barbara Pisker[iD], and Gordan Paun[iD]

Faculty of Tourism and Rural Development, The University of Osijek, Pozega, Croatia
{kdokic,bpisker,gpaun}@ftrr.hr

Abstract. Search engine bias is a reflex of an overall social system intertwined with discriminatory, prejudiced, or biased practices of different social groups. This paper focuses on the question of ethnic-biased search engine results of European women images. The paper aims to examine whether three culturally diverse search engines (Google, Yandex, and Baidu) will result in different nudity scores for nine EU-27 selected women ethnicities.

For the paper, 100 photos of women from nine EU countries were collected using three different search engines. After that, the nudity score was calculated for the 2700 photos, and the results were compared using suitable statistical methods. The results indicate a statistically significant difference between the search engines regarding the nudity score of the collected photos, whereby we can conclude that the results of the Chinese search engine are the most liberal. At the same time, the other two are more conservative.

Keywords: search engine bias · women · culture · NSFW · nudity score

1 Introduction

Search engine bias has provoked many controversies in the ICT sector, especially as it has primarily been understood as a technical issue. As numerous interdisciplinary, especially social sciences research has proven recently, search engine bias is just a reflex of an overall social system, intertwined with different discriminatory, prejudiced, or biased practices of different social groups (in terms of race, ethnicity, gender, minority, culture, religion than the majority) are facing, over spilled from physical to digital society.

Since our research questions aimed to tackle gender bias primarily, we have deepened it by interconnecting it to the context of nudity through different European ethnicity, examining it through three culturally diverse search engine results. Although if we were to count different types of search engine bias found in recent literature, it would pile up to a few dozen, gender bias we found most prominent. Therefore, this research aims to determine whether gender bias (if present) is cross-culturally different when targeting different ethnicities of women in the EU-27 countries.

The research itself tested differences in the perception of European women with nine ethnicities (from the nine largest EU-27 countries) based on photo nudity scores at three culturally different search engines (Google, Yandex, and Baidu). Two research questions were asked:

© IFIP International Federation for Information Processing 2023
Published by Springer Nature Switzerland AG 2023
N. Edelmann et al. (Eds.): ePart 2023, LNCS 14153, pp. 117–130, 2023.
https://doi.org/10.1007/978-3-031-41617-0_8

1. If we analyze each search engine separately, is there a difference between the nudity score of the photographs obtained for the mentioned 9 EU countries?
2. Is there a difference between the nudity score of photographs obtained for a particular country and on different search engines?

The obtained results confirm the author's hypothesis about the existence of gender-ethnically conditioned expansion of the diversity of bias from real society to digital society through biased results we get from search engines.

In presenting the topic frame, this paper comprises five main parts: an introduction, a literature review on relevant recent research and an introspective on the gender-ethnic bias, the concept of nudity and machine learning algorithms for nudity detection, followed by the data and methodology used in research, results and discussion including policy recommendation, and final remarks.

2   Literature Review

2.1   Socially Conditioned Search Engine Algorithms

As with any other human product, search engines also reflect characteristics of our societal sphere (practices, institutions, infrastructure, culture, politics, power, technology, and history), simultaneously projecting social relations and comprehension in the globally digitalized world of today. Therefore, current search engine systems are fed with, use, produce and spread our existing social labels, prejudices and biases constructed on identity, gender, and ethnicity, reflecting our human subjectivity. As noted by Crawford, K. (2021), looking at how classifications are made, we see how technical schemas reinforce existing hierarchies and magnify inequity [1].

Seeing Search Engines as the "epistemic machinery", Knorr Cetina, K. (2007), with technology continuing to permeate every aspect of society, noted its importance in considering how biases and prejudices can be reflected and amplified. At the same time, these biases can be both intentional and unintentional, resulting in discriminatory outcomes that reinforce existing power structures and inequality [2].

Technology's design and development process are not immune to bias and prejudice. The tech industry is often dominated by a narrow socio-demographic group that may not have diverse perspectives or experiences. This can result in a lack of representation and a limited understanding of the needs and perspectives of underrepresented groups, leading to technologies that don't meet broader societal needs or perpetuate harmful racial, ethnical and/or gender stereotypes [3–5].

Recent literature also shows a growing worry that algorithms utilized by advanced tech systems produce discriminatory results, as they are based on data that contains inherent societal biases. Proving propagation of societal gender inequality by Internet search algorithms is found in Vlasceanu M & Amodio, D.M. (2020), confirming how societal disparity, when reflected in internet search algorithms, can prompt human decision-makers to act that maintain and reinforce existing social discriminations and inequalities [6].

Goffman, E. (1976) pointed out that "public pictures" play a significant role in the manifestation and replication of gender-based "social structures of hierarchy or value" [7]. Additionally, recent studies have demonstrated that algorithmic classification systems serve as tools for reproducing and even intensifying overall, broader social biases [8–10].

Search engine algorithms have been widely criticized for being biased towards displaying sexualized and objectifying images of women. One of the most important contributions to this field of research is Safiya Umoja Noble (2018) work exploring how search engine algorithms can perpetuate and reinforce societal biases, particularly concerning gender. One of the key findings is that search engines tend to reinforce and perpetuate gender stereotypes in their image results. Noble highlights the issue of the sexualization of women in search engine image results. When searching for images of women, search engines often prioritize and display sexualized images, reinforcing the idea that women are primarily objects of male desire. This can seriously affect women's self-esteem and comprehension by other different social groups [9].

Similarly, Kate Crawford's (2016) research has revealed that search engines often display objectifying images of women, particularly in relation to sexualized keywords. This can contribute to the objectification and sexualization of women in society [11].

Finally, as noted by Silva, S. & Kenney, M. (2019), without proper mitigation, preexisting societal bias will be embedded in the algorithms that make or structure real-world decisions [12], serving as a perpetuating discrimination engine.

2.2 Nudity as a Cross-Cultural Category

According to Hofstede (2001), cultural differences can be understood by examining the collective programming of the mind that distinguishes one group from another. This concept of culture as "collective programming" provides a powerful tool for comparing national cultures and can serve as a starting point for understanding cultural differences [13]. Cultural differences in attitudes towards nudity reflect how cultural, religious and historical factors shape norms and values related to the human body and sexuality. These differences highlight the diversity of cultural understandings of nudity and how broader cultural, political and historical contexts shape these understandings.

Foucault, M. (1977) explored the role of the body and nudity in the development of modern Western societies as a subjective to social norms and moral values [14]. In his work "The History of Sexuality", Foucault argues that sexuality is not a natural or biological phenomenon but a socially constructed concept that has been heavily influenced by dominant power structures and the institutions that enforce them. He suggests that regulating sexuality through institutions and the state has significantly shaped our understanding of normal or deviant sexual behavior [15].

Nudity as a concept is embedded in a cultural context. It varies across the world's cultural diversity, as described by Elias, N. (1978), who studied the formation of cultural taboos and the construction of modesty norms, which vary across cultures and periods [16]. Cross-cultural studies on nudity often aim to identify patterns and similarities across cultures and highlight the unique ways in which cultural, historical, and social factors shape attitudes towards nudity in different societies [17–19].

Bordo, S. (1993) analyzed how nudity is used to control and regulate bodies, particularly concerning gender, where nudity is often associated with women's bodies and used to reinforce traditional gender norms and sexual objectification [20]. Additionally,

Butler, J. (2005) argues that gender, like nudity, is a highly regulated and policed aspect of social life and that individuals who deviate from dominant norms around gender and sexuality are often subject to stigma and marginalization. By challenging the idea that gender is a natural, biologically determined characteristic, Butler highlights how gender norms and expectations are imposed and maintained through cultural and institutional practices [21].

Western civilization in comprehending nudity has historically been associated with classical Greece and Rome, where nudity was accepted in specific social contexts, such as athletic competitions, and was also depicted in art. However, the rise of Christianity led to a shift towards more modest attitudes towards nudity and the development of modesty norms that continue to shape Western attitudes towards nudity today. Nudity is variously legitimized in contemporary Western culture through the context of representation and placement. Sexualization of the public sphere destabilizes the contexts in which non-sexual nakedness and gazing have been legitimated in modernity [22].

In China, nudity has been historically associated with shame and avoidance and has also been linked to cultural practices such as foot binding and clothing practices. In recent years, however, there has been some liberalization of attitudes towards nudity in China, as seen in the growth of the body-positive movement and increased acceptance of nudity in advertising and media [23].

Russian society's attitudes towards nudity have varied throughout history, from more liberal attitudes during the Soviet era to more modest attitudes in post-Soviet Russia. The role of Russian Orthodox Church, in shaping attitudes towards nudity has played a significant role (similar to the Catholic church in Western civilization) and has been a critical figure in this shift towards contemporary re-traditionalization [24].

3 Research

3.1 Sample

The search engines that will be analyzed were first selected to test the differences in the perception of women of different ethnicities. These were the Google search engine (www.google.com), Yandex search engine (www.yandex.ru) and Baidu search engine (www.baidu.cn). Google search engine is the most used search engine in the United States, Yandex search engine in Russia and Baidu search engine in China. After selecting search engines, the countries to be analyzed were chosen. Nine European Union countries with the most significant number of inhabitants were selected. These are Germany, France, Italy, Spain, Poland, Romania, the Netherlands, Belgium, and the Czech Republic. Considering that all three mentioned search engines search and index the entire Internet space. Our goal was to compare the perception of women in the nine said EU countries, the words "name of the country" (e.g. Germany) and "woman" were entered into the search engines, but in the language of the country in which the search engine is located. In other words, if we wanted to test the perception of German women by the Google search engine, i.e. Internet users who upload photographs with texts about German women in the English language, the words "Germany woman" were entered into the search engine google.com. In case we wanted to test the perception of German women by the search engine Yandex, that is, Internet users who post photographs with texts about women in the Russian language, we entered the words: Германия женщина into the search engine yandex.ru. After the search engine returned images to the default query, the first 100 images for each country and from each search engine were downloaded. In this way, 300 photographs of women were collected for each country. For 9 EU countries, the total number of collected images was 100 images × 3 search engines × 9 countries = 2700 images.

3.2 Method

The assumption is that the collected images can be used as an indicator of the perception of significant content creators on each of the three listed search engines and the algorithms of the search engines themselves. The main problem with this approach is that it is complicated to estimate the ratio of influence of significant content creators in a specific language concerning the algorithm of the search engine being tested. Based on this, it is difficult to make a general conclusion about the entire population of users of a particular search engine in a specific language because it is about three search engines with different and publicly unavailable algorithms for sorting results. Nevertheless, regardless of the mentioned limitations, the described approach indicates different outcomes that were obtained, which could influence the user's perception of the search engines themselves. Figure 1 shows the information flow between a significant content creator (e.g. a journalist of a web portal) who publishes articles and images on a web server. Search engines index these texts and images, and the search engines are then used by users, distributing information that includes the biases of the significant content creator and the specifics of the search engine algorithm being used.

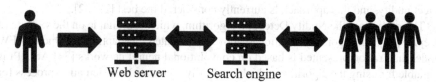

Web server   Search engine

Fig. 1. Information flow (authors)

By reviewing the collected images, it was possible to see the specifics of women from individual countries, so it is straightforward to see the difference between images of women from Spain and images of women from Germany, regardless of the search engine. However, a nudity score was calculated for each image to investigate the previously described perception. A nudity score is usually a number between 0 and 1 that results from a AI algorithm. A value of 1 represents a 100% probability that a human is without clothes in the image, while a value of 0 is obtained if the probability is 0%. As a rule, a result is a number between 0 and 1. An artificial intelligence algorithm was used to get this value, which guaranteed us objectivity and impartiality. The analysis of the obtained values should reveal the prejudices of certain groups and the specificities of specific algorithms of search engines towards women who belong to the mentioned nine EU states or ethnicities.

There are many artificial intelligence algorithms for obtaining a nudity score. Ananthram et al. analyzed ten of the most famous ones for which an Application Programming

Interface (API) is available on the Internet. In the title of Ananthram's paper, the author mentioned the abbreviation NSFW for the algorithms. It is a set of algorithms often used by companies to detect content that is not suitable for use during working hours, and the abbreviation comes from "Not Safe For Work". The authors stated that NSFW algorithms detect five categories of content, namely:

- Explicit Nudity
- Suggestive Nudity
- Porn/sexual act
- Simulated/Animated porn
- Gore/Violence [25]

For this paper, some of the listed categories are unimportant, but they do not affect the result.

There are different approaches to detecting human beings without clothes, and the main goal was to automate this process. Garcia et al. proposed a model based on human skin colour and achieved a precision of 90.33% and an accuracy of 80.23%. They transferred the images to YCbCr space and classified them depending on whether a pixel was skin-coloured [26] Moreira et al. proposed a new dataset of 376,000 images categorized into pornography and non-pornography. The authors used convolutional neural networks, namely Densenet-121, with a batch size of 128 trained with an SGD optimizer and a learning rate of 2^{-8}. The authors state that they achieved an overall accuracy of 97.1% on the combined datasets, and they consider convolutional neural networks to be the best choice for detecting pornography in images [27] At the end of the second decade of this century, a whole series of authors started using convolutional neural networks to detect nudity, and this approach is currently considered the best [29–32].

The paper uses the Nudity Detection algorithm, which is available on the website of DeepAI [33] The algorithm mentioned above was created by adapting the Open NSFW model that Yahoo presented is based on convolutional neural networks [34] An API is available for using the algorithm, which can easily automate the calculation process for many images. A simple program that performed the automation is available at: https://github.com/kristian1971/NSFW.

3.3 Results

Tables 1, 2, and 3 provide descriptive statistical data on the obtained values for the analyzed three search engines. In the columns are the abbreviations of the researched states (ISO 3166), while in the rows are the abbreviations for average (AVR), median (MED), standard deviation (STDV), and maximum (MAX).

Figure 2 shows the average nudity score values for individual countries and search engines, and Fig. 3 shows the median. Figures 4 and 5 show the standard deviation and maximum nudity score for respective countries and search engines.

The first research question was whether there is a difference between the nudity score of the images obtained for the mentioned 9 EU countries if we analyze each search engine separately. The non-parametric Kruskal-Wallis test was chosen, and the hypothesis that there is no difference between countries was set. PSPP GNU 1.6.2-g78a33a software was used for statistical analysis. The results are below in Table 4.

Table 1. Google

	DE	FR	IT	ES	PL	RO	NL	BE	CZ
AVR	0.025	0.031	0.052	0.074	0.023	0.030	0.037	0.032	0.088
MED	0.005	0.004	0.009	0.008	0.002	0.006	0.007	0.002	0.017
STDV	0.065	0.078	0.121	0.160	0.074	0.061	0.087	0.082	0.153
MAX	0.420	0.549	0.944	0.787	0.585	0.420	0.504	0.483	0.654

Table 2. Yandex

	DE	FR	IT	ES	PL	RO	NL	BE	CZ
AVR	0.019	0.016	0.064	0.028	0.040	0.019	0.037	0.015	0.089
MED	0.002	0.002	0.012	0.008	0.006	0.003	0.010	0.002	0.023
STDV	0.067	0.049	0.142	0.068	0.070	0.057	0.066	0.050	0.160
MAX	0.565	0.357	0.808	0.639	0.372	0.397	0.463	0.417	0.891

Table 3. Baidu

	DE	FR	IT	ES	PL	RO	NL	BE	CZ
AVR	0.060	0.042	0.073	0.073	0.058	0.040	0.066	0.052	0.038
MED	0.018	0.009	0.020	0.025	0.012	0.007	0.017	0.012	0.005
STDV	0.123	0.077	0.156	0.132	0.132	0.077	0.138	0.104	0.097
MAX	0.754	0.456	0.938	0.807	0.886	0.379	0.852	0.739	0.597

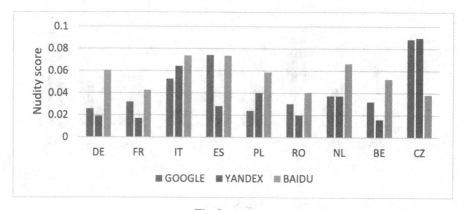

Fig. 2. Average

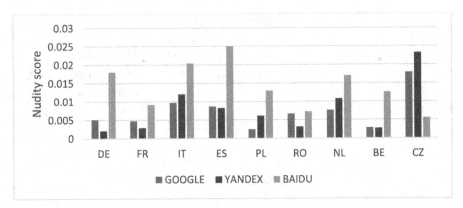

Fig. 3. Median

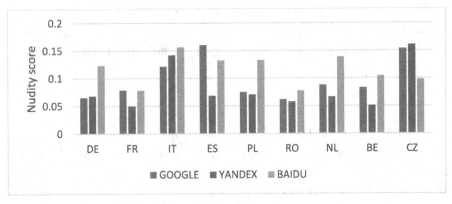

Fig. 4. Stdev

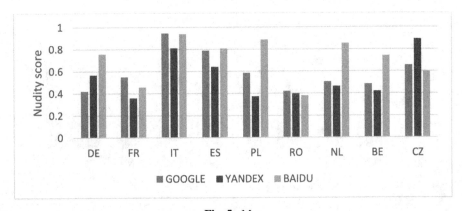

Fig. 5. Max

Table 4. The Kruskal-Wallis test results by search engine

	N	Mean Rank Google	Mean Rank Yandex	Mean Rank Baidu
DE	100	432	336	476
FR	100	416	370	428
IT	100	514	521	502
ES	100	491	490	524
PL	100	347	476	447
RO	100	460	385	407
NL	100	445	520	468
BE	100	389	355	463
CZ	100	557	596	336
Chi-Square		49,52	97,62	36,29
Sign		0,000	0,000	0,000

All three hypotheses were rejected, and the statistical significance was $p < 0.000$, indicating a significant difference in the nudity score of images from nine EU countries on all three search engines.

The second research question is whether there is a difference between the nudity score of images obtained for a particular country and on different search engines. The non-parametric Kruskal-Wallis test was chosen, and the hypothesis was set that there is no difference between the nudity score values of one country for different search engines. The results are shown below for each of the nine states separately.

The Kruskal-Wallis test revealed a statistically significant difference in the level of nudity score in the images of women from **Germany, France,** Spain, **Poland, Romania, Netherlands, Belgium** and **Czechia** on three different search engines (GP_google, n = 100; GP_yandex, n = 100; GP_baidu, n = 100).

The Kruskal-Wallis test did not reveal a statistically significant difference in the level of nudity score in the images of women from **Italy** on three different search engines (GP_google, n = 100; GP_yandex, n = 100; GP_baidu, n = 100).

All results are in Table 5 and Fig. 6.

4 Discussion

The first three hypotheses that there is no significant difference in the nudity score of images from nine EU countries on all three search engines was rejected so that we can comment on the obtained results. Figures 2 and 3 show the averages and median nudity scores for all nine countries and all three search engines. It is interesting that for as many as 7 out of 9 countries, the average nudity score value is the highest for the Baidu search engine, while for the Google search engine, this is the case only for Spain and for the Yandex search engine, this is the case only for the Czech Republic. The median nudity score for even 8 out of 9 countries is the highest for the search engine Baidu, while

Table 5. The Kruskal-Wallis test results by country

	Chi-Square	p	Md Google	Md Yandex	Md Baidu
DE	39.64	0.000	0.005	0.002	**0.018**
FR	16.15	0.000	0.005	0.003	**0.009**
IT	4.58	**0.101**	0.009	0.012	**0.020**
ES	13.84	0.001	0.009	0.008	**0.025**
PL	23.28	0.000	0.002	0.006	**0.013**
RO	10,12	0.006	0.0066	0.0031	**0.0072**
NL	7.51	0.023	0.008	0.011	**0.017**
BE	32.17	0.000	0.003	0.003	**0.013**
CZ	24.56	0.000	0.018	**0.023**	0.006

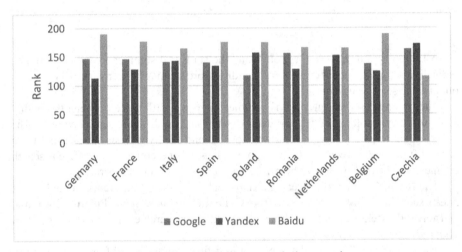

Fig. 6. The Kruskal-Wallis test results by countries

the only exception is Yandex and again only for the Czech Republic. Figures 4 and 5 show standard deviations and maximum nudity scores for all nine countries and all three search engines. In these figures, it can be seen that the highest values belong to the search engine Baidu, namely the five largest maximums and the six most significant standard deviations. This result might be understood through Hofstede's cultural dimensions theory with Russia and China being strongly restrained cultures when compared to the Western civilization [13]. Similarly, as confirmed by Butler (2005) gender norms and expectations are imposed and maintained through cultural practices [21].

The same concept of cross-cultural understanding can be applied to understand the following findings. We state that the Baidu search engine for most analyzed nationalities displays images whose nudity score is significantly higher than other search engines. Interestingly, the only one that stands out is the Czech Republic, for which the Yandex search engine displays images with the highest nudity score values. Of the nine countries listed, only three were once part of the Warsaw Pact, and there were strong political and economic ties between those three countries and Russia. After the dissolution of that alliance, it can be said that the Czech Republic is the most liberal of the three states mentioned. This anomaly indicates the specific attitude of the Russian electronic media towards women in the Czech Republic. These results also confirm how cross-cultural studies on nudity identify patterns and similarities across cultures and highlight the unique ways in which cultural, historical, and social factors shape attitudes towards nudity in different societies [17–19].

The second research question refers to nine countries, so nine hypotheses were set that there is no difference between the nudity score values of one nation for different search engines. All hypotheses were rejected except for Italy. In most cases, the ranking was highest for the search engine Baidu, except for the Czech Republic, where the ranking was again highest for Yandex. As for Italy, there is no statistically significant difference in the nudity score of images from the analyzed search engines.

5   Conclusion

A socio-technical approach to problematizing the topic of bias in search engines has resulted in a deeper understanding of this complex issue by proving how social systems are indirectly reflected in search engine systems, revealing technical results that are often attributed to prejudices, stereotypes or discriminatory practices of the creator himself. Since technology is undoubtedly the driver of social development, primarily because modern development in this field concerns the role of search engines, which acts not only as a performer but also as a master of the mind in various areas of social life, we are undoubtedly the subject of discussion, to examine and test its effectiveness on gender-ethical reasoning, justifications, and non-discriminatory practices in terms of universal human rights. The main goal of this paper is to point out this specific question about potentially ethnically biased search engine algorithms that manifest as different levels of nudity in European women images.

The main observations and findings presented in this paper are as follows:

1. There is a statistically significant difference in the nudity score of photos of women from nine EU countries on all three search engines individually.
2. There is a statistically significant difference in the nudity score of photos of women from nine EU countries when we compare the results for each country individually on all three search engines, except Italy, where there is no statistically significant difference.
3. The most liberal search engine is the Chinese Baidu, except in the case of the Czech Republic, where the highest average nudity score is obtained from the Yandex search engine.

Regarding theoretical contributions of this research has confirmed classical taught on cross-cultural societal influence and differences shape our world, even though transferred to digital sphere. Methodologically the research has indicated and presented relevant and reliable, easily repeated methods for further and different cross-cultural testing of Search

Engine bias. Practically, while proving Search Engine bias it emphasized the importance of overcoming and mitigating strategies development and considerations, deploying bias free future Search Engines. The overall relevance of this research, although narrowed down to search engine bias, can serve as a concept model in understanding socio-technical biases in our different societal spheres: economy, policy, government and medicine. Especially with the initiation of artificial intelligence in different decision making processes (from testing and hiring to individually fitted medical therapy). This research therefore underlines an undeniable need for more transparent, understandable, inclusive and balanced technology approach fitted to the differences we share globally.

We see the limitations of the paper in the fact that it is unclear where ethnic bias arises. The source can be from significant content creators (e.g. journalists of a web portal) or search engine algorithms that are inaccessible to us. The assumption is that the search engine algorithms do not influence the results depending on the country, so we tend to believe that it is the authors. Furthermore, the fact is that the obtained average nudity score values are low, and the distribution is stretched from the vertical axis to the right, with the maximum being along the vertical axis. This forced us to use non-parametric tests that detect a difference that may not be noticeable to the average person in the obtained photos of women. In addition, only one nudity detection system was used in the paper, and several are available, the inclusion of which would undoubtedly contribute to objectivity.

Further research could include a more significant number of photos of women and a comparison of the results when searching on all three search engines and in all three languages. In doing so, they would get nine combinations instead of the three combinations used in this paper.

References

1. Crawford, K.: The atlas of AI: Power, Politics, and the Planetary Costs of Artificial Intelligence. Yale University Press, New Haven (2021)
2. Cetina, K.K.: Culture in global knowledge societies: knowledge cultures and epistemic cultures. Interdisc. Sci. Rev. 32, 361–375 (2007)
3. Savoldi, B., Gaido, M., Bentivogli, L., Negri, M., Turchi, M.: Gender bias in machine translation. Trans. Assoc. Comput. Linguist. 9, 845–874 (2021)
4. Jones, J.J., Amin, M.R., Kim, J., Skiena, S.: Stereotypical gender associations in language have decreased over time. Sociol. Sci. 7, 1–35 (2020)
5. Draude, C., Klumbyte, G., Lücking, P., Treusch, P.: Situated algorithms: a sociotechnical systemic approach to bias. Online Inf. Rev. 44, 325–342 (2020)
6. Vlasceanu, M., Amodio, D.M.: Propagation of societal gender inequality by internet search algorithms. Proc. Natl. Acad. Sci. 119, e2204529119 (2022)
7. Goffman, E., Goffman, E.: Gender display. In: Gender Advertisements, pp. 1–9 (1976)
8. Friedman, B., Nissenbaum, H.: Bias in computer systems. ACM Trans. Inf. Syst. (TOIS) 14, 330–347 (1996)
9. Noble, S.U.: Algorithms of oppression. In: Algorithms of Oppression, New York University Press, New York (2018)
10. Schwemmer, C., Knight, C., Bello-Pardo, E.D., Oklobdzija, S., Schoonvelde, M., Lockhart, J.W.: Diagnosing gender bias in image recognition systems. Socius 6, 2378023120967171 (2020)

11. Crawford, K.: Can an algorithm be agonistic? Ten scenes from life in calculated publics. Sci. Technol. Hum. Values **41**, 77–92 (2016)
12. Silva, S., Kenney, M.: Algorithms, platforms, and ethnic bias. Commun. ACM **62**, 37–39 (2019)
13. Hofstede, G.H., Hofstede, G.: Culture's consequences: comparing values, behaviors, institutions and organizations across nations, sage (2001)
14. Foucault, M.: "panopticism" from" discipline & punish: The birth of the prison. Race/Ethnicity: Multidisciplinary Global Contexts **2**, 1–12 (2008)
15. Foucault, M.: The History of Sexuality: The Use of Pleasure, vol. 2. Vintage, New York (2012)
16. Elias, N.: The Civilizing Process: The History of Manners: Sociogenetic and Psychogenetic Investigations, Trans. Edmund Jephcott. Basil Blackwell, Oxford (1978)
17. Smith, H.W.: A modest test of cross-cultural differences in sexual modesty, embarrassment and self-disclosure. Qual. Sociol. **3**, 223–241 (1980)
18. Hetsroni, A.: Sexual content on mainstream TV advertising: a cross-cultural comparison. Sex Roles **57**, 201–210 (2007)
19. Huang, Y., Lowry, D.T.: An analysis of nudity in Chinese magazine advertising: examining gender, racial and brand differences. Sex Roles **66**, 440–452 (2012)
20. Bordo, S.: Unbearable weight: femisisn, western culture, and the body (1993)
21. Butler, J.: Gender trouble: feminism and the subversion of identity GT. Pol. Theory **4**, 4–24 (2005)
22. Cover, R.: The naked subject: nudity, context and sexualization in contemporary culture. Body Soc. **9**, 53–72 (2003)
23. Zito, A., Barlow, T.E., Barlow, T.E.: Body, Subject, and Power in China. University of Chicago Press, Chicago (1994)
24. Barta, P.I.: Gender and sexuality in Russian civilization. Routledge, London (2013)
25. Ananthram, A.: Comparison of the best NSFW Image Moderation APIs 2018. Towards Data Science, 22 November 2018. https://towardsdatascience.com/comparison-of-the-best-nsfw-image-moderation-apis-2018-84be8da65303. Accessed 7 Jan 2023
26. Garcia, M.B., Revano, T.F., Habal, B.G.M., Contreras, J.O., Enriquez, J.B.R.: A pornographic image and video filtering application using optimized nudity recognition and detection algorithm. In: 2018 IEEE 10th International Conference on Humanoid, Nanotechnology, Information Technology, Communication and Control, Environment and Management (HNICEM) (2018)
27. Moreira, D.C., Pereira, E.T., Alvarez, M.: PEDA 376K: a novel dataset for deep-learning based porn-detectors. In: 2020 International Joint Conference on Neural Networks (IJCNN) (2020)
28. Huang, Y., Kong, A.W.K.: Using a CNN ensemble for detecting pornographic and upskirt images. In: 2016 IEEE 8th International Conference on Biometrics Theory, Applications and Systems (BTAS) (2016)
29. Li, K., Xing, J., Li, B., Hu, W.: Bootstrapping deep feature hierarchy for pornographic image recognition. In: 2016 IEEE International Conference on Image Processing (ICIP) (2016)
30. Ou, X., Ling, H., Yu, H., Li, P., Zou, F., Liu, S.: Adult image and video recognition by a deep multicontext network and fine-to-coarse strategy. ACM Trans. Intell. Syst. Technol. (TIST) **8**, 1–25 (2017)
31. Surinta, O., Khamket, T.: Recognizing pornographic images using deep convolutional neural networks. In: 2019 Joint International Conference on Digital Arts, Media and Technology with ECTI Northern Section Conference on Electrical, Electronics, Computer and Telecommunications Engineering (ECTI DAMT-NCON) (2019)
32. Zhou, K., Zhuo, L., Geng, Z., Zhang, J., Li, X.G.: Convolutional neural networks based pornographic image classification. In: 2016 IEEE Second International Conference on Multimedia Big Data (BigMM) (2016)

33. Kumar Thakur, R.: Detect Nudes Using Python Programming and Deep AI, Medium, 10 April 2022. https://medium.com/geekculture/detect-nudes-using-python-programming-and-deep-ai-a9be69b2e9af. Accessed 4 Jan 2023
34. Woodie, A.: Yahoo Shares Algorithm for Identifying 'NSFW' Images, datanami, 3 Oct 2016. https://www.datanami.com/2016/10/03/yahoo-shares-algorithm-identifying-nsfw-images/. Accessed 7 Jan 2023

From Integration to Data Sharing - How Developers Subvert the Public Sector

Daniel Rudmark[1,2]([envelope]) [iD] and Antonio Molin[1,3] [iD]

[1] Swedish Center for Digital Innovation, University of Gothenburg, Gothenburg, Sweden
daniel.rudmark@ait.gu.se
[2] The Swedish National Road and Transport Institute, Gothenburg, Sweden
[3] The Swedish Social Insurance Agency, Gothenburg, Sweden

Abstract. This paper explores how organizations expand data-sharing capabilities beyond the loci of emergence. This inquiry was ignited from an observation that external developer practices had subverted a public sector organization into developing transformational data-sharing capabilities that effectively replaced existing integration practices within the agency. To detail and explain how the administration was able to draw on the practices and platforms established for an external context in an organizational (internal) context, we analyzed our empirical dataset using dynamic capabilities theory. By unpacking enabling microfoundations and overarching capabilities, we could explain our observations and put forward six microfoundations that underpin three data-sharing capabilities.

Keywords: Dynamic Capabilities · Data Sharing · Platform Ecosystems · Digital Government

1 Introduction

Digital transformation in the public sector can help optimize resource allocation, enhance efficiency, and provide citizen with better digital services. However, successful implementation hinges on fostering collaborative relationships with the surrounding society. By partnering with e.g., academia, private entities, and engaged citizens, public sector organizations can benefit from a wealth of diverse skills and knowledge [1], which in turn spurs innovation through the exchange of ideas and technologies [2]. In essence, actively involving citizens and civil society organizations ensures that public sector initiatives are tailored to people's needs and expectations [3], can result in increased satisfaction and trust in government institutions.

One such opportunity for increased collaboration emerges from the proliferation of digital platforms. Here, the commercial sector increasingly leverages digital platforms for their value propositions, to the extent that 7 of the top 10 valued S&P 500 firms had platforms at the very core of their business operations in January 2022. A large part of this success is due to collaboration with actors outside these platforms. By enticing

© IFIP International Federation for Information Processing 2023
Published by Springer Nature Switzerland AG 2023
N. Edelmann et al. (Eds.): ePart 2023, LNCS 14153, pp. 131–147, 2023.
https://doi.org/10.1007/978-3-031-41617-0_9

large numbers of external developers to complement platforms with, e.g., apps and plug-ins, platform companies can offer customers unprecedented service bundles. Given that innovation activities within platform ecosystems this way increasingly moves outside firm boundaries, it has been demonstrated that developers, in practice, *invert* the entire innovation locus of these platform companies [4].

However, the public sector still experiences difficulties reaping the benefits from this societal digital transformation. Platform ecosystems thrive on agility, flexibility, and rapid adaptation to changing market conditions [5]. In contrast, government agencies are often constrained by bureaucracy and risk aversion [6]. This incongruence between logics can create barriers to effective collaboration and leveraging these ecosystems for public service delivery [7].

In the longitudinal study presented in this paper, we initially found such an inability to act on possible collaboration opportunities regarding app development. However, after recognizing the benefits of and aligning with data-sharing requirements from platform ecosystem collaborations, we also noted how developer expectations and their associated data-sharing practices fundamentally changed how the administration exchanged other data within and across agency boundaries. Put differently; these practices by external developers became so influential that they in practice *subverted* the existing integration paradigm within the administration. Given the importance of public sector data sharing capabilities to effectively enable data-driven decision-making, and cross-agency collaboration, this puzzling observation ignited a more careful investigation of what capabilities and underlying microfoundations enabled this surprising transition. To this end, this research explored the following research question:

How do organizations acquire data-sharing capabilities by engaging with external developers in platform ecosystems?

The rest of the paper is structured as follows. We continue by accounting for how systems integration in a platform world may need new approaches, followed by outlining this paper's theoretical framework, *dynamic capabilities*, and the theory's relation to causation and effectuation. We next account for this research's methods and analytical procedures and continue by describing the microfoundations that underpinned this successful transition. We continue the paper by discussing our findings and putting forward three salient dynamic capabilities that enabled the STA to navigate and benefit from the new order. We end this paper by offering implications for practice and answering our research question.

2 Systems Integration in a Digital Platform Age

For public sector information systems, systems integration is critical within and across agency boundaries [8–10]. Within this stream of research, there is an underlying assumption that the parties integrating are known and that the ontological assumptions (e.g., how to represent a particular aspect of reality [11]) underlying the integration are negotiated [12].

Given the digital transformation and platformization of society, these underlying assumptions regarding integration within and across agencies have become increasingly challenged [13]. Integration in platform ecosystems instead rests on an assumption of stable interfaces at the core enabling a large variety in the periphery (such as apps) [14]. To foster this variety, platforms must be able to scale new complements (and their necessary integrations) without marginal costs for the platform owner [15, 16]. To achieve such scaling and allow for seamless changes in complements, digital platforms are often designed to minimize dependencies between the platform and its complementors [17]. Here, a core means is keeping the crossing point interfaces as "thin" as possible [18, 19]. This design decision, e.g., entails minimizing data structures marshaled through the interface, thereby lowering dependencies, complexity, and entry barriers for new platform entrants [20].

In light of these developments, some authors suggest that data sharing may be a more appropriate term to signify integration in platform ecosystems. While this term lacks a generally agreed definition, it typically entails offering access to data following formal or informal principles [21]. This way, integration can be achieved as the parties understand the principles rather than through idiosyncratic bilateral agreements. To better understand how organizations develop data sharing capabilities, we next turn to dynamic capabilities, the theoretical device used in this paper's analysis to more precisely understand how the STA adapted and reconfigured its resources in response to external developer practices.

3 Executing Dynamic Capabilities

Dynamic capabilities are an organizational theory used to describe, analyze, and understand how organizations strategically employ their resources, assets, and differentiators to create advantages in uncertain, and dynamic markets [22]. The turbulent environments addressed by dynamic capabilities are, however, not exclusively shaping private enterprises but also the environments of the public sector [23, 24]. The fierce pace of change, of digitalization, including platformization, require the public sector to employ dynamic capabilities, although not driven by the need for competitiveness but the competition-related value ideals of efficiency, effectiveness, and service quality [25–27] essential to enabling trust in government institutions.

To grapple with high degrees of change and uncertainty, organizations can execute dynamic capabilities to sense openings and threats, seize opportunities, and transform these into new knowledge and practices [28]. To frame and describe how dynamic capabilities are enacted, *microfoundations* are typically used in relation to the higher(top) order dynamic capabilities [28, 29].

Dynamic capabilities extend the resource-based view (RBV) by theorizing how firms create and maintain advantages through the internal configuration, development, and deployment of resources [30]. A core tenet of RBV and, by extension, dynamic capabilities, is the assumption of rationality of strategists, managers, and decision-makers [22]. However, under conditions of high degrees of uncertainty, ad-hoc actions and experimentation may prove more relevant to competitive advantages than the rational configuration of resources and capability development [31]. In such turbulent environments, the capacity for imagination, acting on opportunities, navigating networks, adaptive governance, and distributed decision-making [31, 32] may prove to be a more durable fabric of organizational microfoundations. Given that both these seemingly contradictory perspectives have a bearing on dynamic capabilities, we next turn to how microfoundations can be executed.

In a recent effort to reconcile these conflicting perspectives, [33], based on effectuation theory [34], suggests that decision-making logic perspectives complement the understanding of how competitive advantages are created and opportunities sized through the execution of dynamic capabilities. Here, causally executed processes refer to taking "a particular effect as given and focus on selecting between means to create that effect" [34, p. 245]. Conversely, effectually executed processes instead refer to assuming "a set of means as given and focus on selecting between possible effects that can be created with that set of means" [34, p. 245]. In dynamic multi-stakeholder environments, effectual decision-making manifests through means-driven action, reliance on networks and strategic alliances, affordable loss, and leveraging contingencies. In this sense, effectuation contrasts causal decision-making that emanates from goal-driven action, reliance on competitive analysis, expected returns, and exploitation of pre-existing knowledge [33, 34].

4 Method

This research is based on a 12-year (2010–2022) case study [35] at the Swedish Transport Administration (STA). The Swedish Transport Administration (STA) is the national authority responsible for the long-term planning and ongoing maintenance of the national road and rail transport system throughout Sweden. This responsibility includes building, operating, and maintaining public roads and railways and the associated information services. Following the increasing digitalization of transportation, the administration currently spends some 2,7 billion SEK on IT and employs more than 1500 employees and contractors for IT.

4.1 Data Collection

Data were collected throughout our long-term engagement with the STA and included formal interviews, participation in meetings and workshops, and documents such as emails and internal investigations. All interviews and workshops were transcribed verbatim, whereas meetings were documented through notes. An overview of the data can be found in Table 1.

4.2 Data Analysis

Our approach to data analysis was abductive, an approach that is fruitful when researchers encounter phenomena that are at odds with current theorizing [36] and pursues the analysis to resolve this theoretical surprise. More specifically, we relied on systematic combining [35] to arrive at our findings. Systematic combining uses two overarching processes, matching and direction/redirection. Matching refers to an ongoing analysis that iterates between relevant theoretical concepts and the empirical material to facilitate the development of the theoretical framework and the case in tandem. Direction and redirection refer to data collection and analysis as directed by the evolving theoretical framework but simultaneously open for redirection triggered by the inclusion of alternative empirical material from the case setting. Systematic combining thus enacts

Table 1. Data Collection

Data source	Data collected	Content of data
Interviews the STA	N = 12, Σ = 846 min	Interviews with platform team members, strategists, and managers. Focusing on critical events and rationales from the platform team, current strategies, and necessary strategic movement with the STA strategists and managers
External API user interviews	N = 35, Σ = 1460 min	Interviews with external developers, both early scrapers, later entrants, and train operator representatives. Focusing on development motivation, app functionalities, and feedback on platform capabilities
Workshops	N = 7, Σ = 2190 min	Workshop (N = 1) with developers and representatives from the STA on future platform designs and internal workshops (N = 6) within the STA on how to utilize platform learnings in new ways
Meetings	N = 52, Σ = 2530 min	Meetings with the platform team during the design phase. Covering status updates and detailed design decisions. Informal meetings with platform team members after the launch of the external platform
Emails	N = 8	Emails with external developers and platform team members, following up interviews, typically asking for clarification of responses during data analysis
Online forum discussion posts	N = 62	Discussion between developers and representatives on the API design of the STA platform
Reports	N = 2	One report investigating the architecture and interface design of an external production platform. One report examined whether DataCache could be an official external integration platform
Platform usage data	N = 2	One spreadsheet with longitudinal external usage data. One sheet with data on internal usage of DataCache

"a nonlinear, path-dependent process of combining efforts with the ultimate objective of matching theory and reality." [35, p. 556].

As part of a follow-up study of the external platform initiative, we encountered a surprising finding. It is well established that external developers can "invert" the innovation locus outside the firm [4]. However, to the best of our knowledge, we lack reports of how such massive external developer activity can substantially affect *internal* practices. Given this surprise, we started to reanalyze and partly collect new empirical material. As a first step, and using a previous analysis in Atlas.ti, the first author wrote a more extended case narrative (word count = 9954) encompassing key events and supporting evidence of the surprise. Using this narrative, we created a timeline of events. During this matching process, we, in addition to recognizing the developer actions, also acknowledged the actions of the STA to accommodate the change. When revisiting the underlying empirical material, we thus redirected our study to use *dynamic capabilities theory* [22, 28]. We carefully selected this framework as analytical device given its frequent use in explaining throughgoing organizational transformations triggered by novel and external change agents [37], and where the dynamic capabilities are retrospectively discovered [38].

As we matched the data with the sensing, seizing, and transforming concepts, we also noted a significant shift of execution for the actions within the STA after the external platforms were deployed. After iterating our empirical data with the dynamic capabilities' literature, we found that *effectuation theory* [33, 34] helped us explain this execution shift. Using these theoretical constructs, we entered an intense matching process where we iterated between theory and data to elicit the microfoundations and capabilities similar those presented in this research. However, since these capabilities had been developed using existing data collected for slightly different purposes, we engaged in additional data collection. This data included interviews explicitly focusing on the concepts in this paper alongside internal platform usage data, emails, and systems documentation. Using this data, we arrived at the final findings presented in this paper.

5   Results

In 2010, the STA did not publish data for third-party developers, but data was nonetheless scraped from the STA's website, using its HTML code and javascript interfaces. This real-time data was reprocessed and conveyed through smartphone apps developed by independent developers, primarily driven by self-experienced frustration with the STA's inability to utilize this new technology. A handful of the applications gained wide popularity with hundreds of thousands of downloads in application marketplaces.

5.1   Microfoundation: Platform Affordances Co-exploration (2010–2011)

While the services were largely ignored at the official level, two STA officials were in contact with developers and noted how the data interfaces used internally by these developers were simpler than those used at the time by the STA to exchange data with externals. The type of data infrastructure requested by developers (simple JSON objects, marshaled through REST APIs) to disseminate real-time data was not possible to develop within the boundaries of current strategies within the STA. Consequently, a group of strategists, led by the manager of traffic information within the STA, chose to engage in an R&D project to investigate how data could be published. On 2012-04-19, the project held a joint workshop summoning representatives from STA, developers of unsanctioned applications, and other R&D project team members. This workshop's purpose was to bring different stakeholders together for the first time and jointly explore what these developers needed, as commented by a strategist at the STA:

> *We must understand what needs developers have regarding things like formats, delivery qualities, and content. We also need to know why they need this to understand the value for us of actually delivering it in a better way, not just that developers want something free of charge.*

During the workshop, several insights and tentative design hypotheses were formed. First, the STA participants were surprised by the external developers' interest, commitment, and seriousness. Second, developers voiced the need for two main platform capabilities. The first included a simple interface for commonly used functionality (like getting all departures from a station), and the second the possibility to explore innovations beyond such common functionality. Given these identified opportunities the STA gave the go-ahead to start designing and deploying a live, time-boxed API solution.

5.2 Microfoundation: Ecosystem Alignment Experimentation (2012–2013)

The R&D project used an external API platform that provided two interfaces, TrainInfo (implementing common use cases) and TrainExport (channeling all data points), facing third-party developers while being decoupled from the STAs underlying system feeding the API with data. The purpose of the project was to explore more precisely what type of solutions these developers could develop, and the necessary design of the platform boundary resources to achieve adoption. Consequently, it was instrumental in having the developers invest a similar amount of time, energy, and commitment into their work as they would if the platform were to be sustained over time. For this reason, the APIs came with a time constraint that the STA assessed as sufficiently distant for developers to consider constraining (some 12 months after launch). The STA first put the specification solution openly for discussion. The platform was subsequently released in October 2012 where anyone could register for API access. In three months, 59 developers had registered for this opportunity, and 17 agreed to be interviewed.

To summarize developers' impressions, users that focused on the interface on common functionality (TrainInfo) found it utile. Most of these developers were new to the railway domain but nonetheless voiced that they could use the API to match their needs. Consequently, when asked to summarize their overall views from using the API, all users of TrainInfo echoed a pleasant experience, as exemplified by one developer:

> *I'm positively surprised; I think TrainInfo works very well; it was straightforward to get started. Two words describe it well, quick and easy...if you only have a little knowledge of the world of APIs and development, the rest will follow quickly.*
> Developer B4

However, third-party developers that had used the API that provided access to all data points conveyed a more complex picture. Especially users (two developers) that had existing, popular applications based on scraping instead expressed disappointment and had, for this reason, stuck with unsanctioned data access, as commented by one developer of a popular application:

No, I won't stop scraping, and that's mostly because I see no reason to, "if it ain't broken, don't fix it," or something like that. There is nothing there that attracts me; I will stick to the current solution as long as there is no real reason to switch.

Developer B14

5.3 Microfoundation: Autonomous Architectural Reconciliation (2014–2015)

The previous large-scale pilot project had been executed to inform a possible operational design without committing to a permanent operation. Developers were, however, still dissatisfied with aspects of the platform functionality, most notably the possibility only to retrieve records changed since the previous request. Nonetheless, the overall outcome of the trial convinced STA to create a more persistent solution, given the interest that the trial had rendered (as more developers enrolled in this trial in just two months, compared to a 5+year road data sharing program). As a result, a new official decision to release train data openly was made by the STA's director of the Business Area Society in early 2013.

A critical decision that remained for the STA was how data should be formatted and released to the developers. To this end, the STA undertook an internal investigation that was used to detail the official decision above. This investigation recommended the STA to depart from the position of only using complex industry standards and instead use a tailored, simplified interface matching developer needs. As the investigation intentionally left more detailed design decisions open, it became the platform team's mission to decide on the precise interface design.

To this end, the platform team embarked on designing a platform that could cater to these needs. In doing so, they chose to substantially re-engineer an existing internal data lake platform that contained core functionality but required substantial modification to meet developer needs. Here, they re-engineered an existing platform's internal query language for reduced redundancy, syntax strictness, data model simplification, and increased congruence between different data models to facilitate more advanced data retrieval. In addition, the new platform included coordinates requested by developers (WGS84), a console for experimenting with queries, example queries for common use cases (the common functionality), simple registration procedures, and the possibility to retrieve records that changed after any given time.

The platform was launched on 2014–03-18, and 12 developers that used the new platform were interviewed. All echoed a positive experience when asked to summarize their experience, as exemplified by a developer lacking previous experience in the railway domain:

Definitely a nice surprise. It wasn't, how do you put it, it wasn't my perception of what the STA was doing. So that it actually exists made me very pleasantly surprised. The API meets my needs. Developer R10

The platform quickly got traction and currently has some 6000 registered developers. In addition, there are presently more external than internal API calls (some 100 000 000 external calls/month vs. 60 000 000 internal calls/month).

5.4   Microfoundation: Exploiting Data-Sharing Opportunities (2016)

This external uptake by developers was both swift and unexpectedly large. In addition, the R&D project evaluations showed that the platform had reached new types of developers new to the railway domain. When reflecting on the differences between the approach in the R&D project and other forms of integration efforts within the agency, the systems architect of DataCache accredited this successful uptake to the platform design process, which was radically different from typical internal projects:

> Usually, you do something for your colleague; in the next room, you typically just meet that need. You get it to work, and when another need shows up, you'll just through something together for that. When we designed the open API platform, we collaborated closely [with developers] to resolve things like "How can we understand the needs of third-party developers? How can we make things easier for them?"

Following this line of argument, the platform team started to use the term "data sharing" rather than "systems integration" to characterize the platform affordances developed in the Open API platform. This notion pertained to the experiences gained within the R&D project, where data should be offered on the premise that you should not assume to know who and why that will integrate with your data, as explained by the integration platform owner:

> I'm starting to dislike the word integration in general. In integration, you have a system A that should talk to system B, and then you create a channel in between, and these two starts talking to each other. We who oversee this [Open API] platform instead think that [...] those who own the information in system A should share this data. We know that system B needs it, but we also think that more people may need this data, so we have to think carefully when we publish the data so that it is useful.

Because of this line of thinking, the platform team hypothesized that the type of uptake effect, as with the Open API, also could be achieved in other, less open initiatives. One such existing initiative within the STA concerned data exchanges with train operators. The STA is essentially a railway service provider for train operators, and in this capacity, it needs to exchange a wide variety of railway data (not shareable with the public). The operators had long been exchanging data through an archaic interface planned to be replaced by newer technologies. Operators had been frustrated by both the complexity and questionable quality of the interface, and the platform team hypothesized that this dissatisfaction could be ameliorated by deploying a separate instance the open API platform.

5.5   Microfoundation: Ecosystem Logic Carryover (2016–2017)

The train operators were largely positive about migrating to functionality similar to the Open API platform. Some operators had already been starting to use publicly available data from the Open API for commercial reasons and found it more pertinent than the existing technology, as commented by an operator developer:

To use [the old interface], there was a contact person from whom you received information, and starting there, you would begin to figure out how [the old interface] worked in a rather undocumented way. The data that the STA delivers [the old interface], it is not good because you cannot connect with other datasets, but [with the open API], you can easily throw together different datasets and, for instance, get pretty nice visualizations.

Consequently, the platform team decided to reuse the interface standards developed with external developers. This approach, however, required collaborative rearchitecting of the data to be shared, as explained by the system architect:

And then [the railway team] came dragging with an extremely complicated data model with obscure coordinates, strange field names, and whatnot, and then we say, "We cannot publish this; no one will ever be able to use it." We must start by knocking together an understandable data model, and once that is done, we say, "And now it's time to document it," and they say, "What – must we also document this?" and we say "Yes." So, we were a little tough on the internal client, but that is all to provide a great experience for those developers who will be using the data.

For the Open API platform to take this role of a new official data exchange platform for train operators, however, the platform needed to be augmented with additional capabilities. Compared to the Open API, access to the data exchanges within the railway value chain was restricted, and access rights to data needed to be based on predefined credentials. To this end, the platform team developed access rights capabilities, effectively enabling the platform to assign granular access to data items depending on roles. In addition, the platform needed to be augmented with functionality to transfer data across network domains. This improvement was necessary as some railway data shared with the STA should only be visible in the restricted domains (e.g., when a train gets the formal signal to depart), whereas others should be publicly available (such as the estimated time of departure). Given that the platform not only served open data it was renamed "DataCache."

5.6   Microfoundation: Leveraging Internal Network Effects (2018–2022)

In 2018 the STA engaged in a central investigation scrutinizing how data more broadly should be exchanged with external actors. This work was instigated because the STA had several technology platforms for this very purpose (DataCache being one). The investigation assessed the different platforms and finally arrived at recommending DataCache as the preferred solution and to replace all other data exchange initiatives within the administration. Following this recommendation, IT management decided to favor the investigation primarily as the platform had proven to establish cost-efficient data exchanges. As a result of this decision, the platform started to cater to a broader array of external data exchanges, including sensor data from vehicle fleets on the road and road survey data from contractors. However, this increased use of DataCache also entailed a surprising side effect - that DataCache organically became extensively used also for purely internal projects, as explained by a platform team member:

*People in the corridors are saying, "We heard rumors about this API platform,"
and then, "we would also like to use it." So, it's all based on mouth-to-mouth,
something like" This is something good, this is something we'd like to use." And
then, if anyone has the need to publish data in any way, we can say to them, "we'll
solve all your problems." They get so incredibly much free. They come to us and
say, "Here's our data," then we do our magic, and all of a sudden, they have a
service up and running in a couple of days that they previously estimated would
take half a year to build.*

The significantly improved integration time was attributed to having carried over
the ecosystem logic used with external developers, into the STA, focusing on the data-
sharing principles (instead of integration). When a team used DataCache for internal
integration efforts, they used a separate instance of the platform where the team could
utilize the same user experience as external developers for the Open API. This approach
included developer registration procedures, thorough documentation, test consoles, and
example queries. However, the increased utilization of DataCache also entailed a massive
increase of datasets being marshaled through the platform. To illustrate, the number of
datasets in different instances of DataCache, increased from 20 in early 2021 to 350
in late 2022. Given this increase, the platform team could no longer oversee all data
conversion from internal systems to DataCache. Instead, teams wanting to publish data
through DataCache needed to create their own schemas, using guiding templates as a
measure to ensure continued data-sharing qualities. Moreover, as additional datasets
were integrated internally through DataCache, the platform (and the team) became an
internal competitor to the official integration team at the STA. To mitigate this situation,
STA IT management moved the team from the business area Traffic to the internal IT
division. Nonetheless, this move was rife with tensions, as explained by the DataCache
product owner:

*And after this move, the [DataCache] team became a bit of a cuckoo in the nest
because we're starting to realize [that] we almost did a kind of hostile takeover
of the existing integration organization. Because all of a sudden, no cases were
coming into their mailbox because we [the DataCache team] had taken all the
internal customers.*

In September 2022, the STA IT manager in charge of integration within the agency
thus decided to phase out the existing integration department and associated technology.
Instead, integration efforts within the firm should be using the DataCache platform,
effectively making DataCache also the internal integration platform.

6   Discussion

This paper explores how external developers not only can invert an organization's innovation locus [39] p. In addition, we noted how these developers, in practice, subverted
an organization into developing internal data-sharing capabilities, effectively replacing
existing integration practices. To resolve this empirical mystery, we analyzed the STA's
development through a lens of dynamic capabilities [22, 28] and paid specific attention

to its microfoundations. Our analysis acknowledged the importance of allowing these microfoundations to be executed from opposite vantage points, i.e., effectually and causally [33, 34]. A schematic picture of the identified micro-foundations and overarching capabilities can be found in Fig. 1.

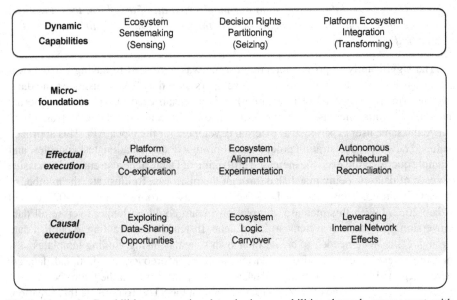

Fig. 1. Dynamic Capabilities to acquire data-sharing capabilities through engagement with developers in platform ecosystems.

6.1 Dynamic Capability: Ecosystem Sensemaking

Public agencies need to develop capabilities to help identify strategic redirections under high uncertainty and insufficient knowledge [31]. In our empirical dataset, we thus noted how the STA engaged in *effectually* executed sensing. During the first sequence of events (2010–2012), the STA successfully captured developer cues rooted in unsolicited external development. At the core of this microfoundation, co-exploring platform affordances were acknowledging the prevailing practices and conventions among external developers were substantially different from existing boundary designs at the STA. However, since the STA lacked knowledge of unpacking developer needs, they engaged in joint, imaginative work with external developers where potential platform configurations based on STA resources were being co-explored. Emerging from these collaborative explorations were requirements such as minimum dependencies through thin interfaces [18, 19], arms-length relationships [5], and the opportunity to both reuse proven solutions as well as explore new territory.

However, the microfoundations enacted to identify possible internal uses of the open API platform were instead causally executed. Here, the platform team predicted that data sharing could also be successfully employed in less open contexts to scale and decouple the integration of systems. Consequently, when the STA chose to replace its data-sharing technology with railway operators, the platform team offered its open API platform. In the light of this analysis, we denote the *causal* activities underpinning sensing exploiting data-sharing opportunities.

Taken together, these two microfoundations make up the organization-wide capability ecosystem sensemaking that we propose can help the public sector to sense data-sharing capabilities from external developers. This capability encompasses how the public sector can sense unexpected uncertainty (as in the effectual collaborative sensemaking of unsolicited app development) and recognizable uncertainties (as when the platform team conjectured that the open API platform could be used to replace the railway data exchange platform).

6.2 Dynamic Capability: Decision Rights Partitioning

A core capability that helped the STA to manage the transformation concerned their ability to arrive at an efficient separation of concerns. To seize the opportunity external developers presented, the STA needed to efficiently distribute the work between the platform core and its complements. However, given the agency's lack of experience with platform ecosystems, they first engaged in an effectual execution of this seizing. Through the microfoundation ecosystem alignment experimentation, the STA iteratively materialized the emerging platform design together with developers without committing to deploying a production platform. This way, the STA could carve out important material platform characteristics, like enabling common use cases as well as more explorative work. Moreover, these experiments convinced the STA to ingest specific requested functionality into the platform core (detecting changes since the last request) and the possibility to easily mix and match data from various objects.

When the STA sought to seize opportunities for more restricted data exchanges, they instead meticulously imposed external developer preferences through a causally executed microfoundation that we label ecosystem logic carryover[1]. Here the platform team used the experiences gained from interactions with external developers. Based on these experiences, they predicted efficiency gains if the same division of work was applied in other data-sharing scenarios that involved the STA (by gatekeeping data publishing in the platform through high publishing standards for the data model). Our analysis exhibited this microfoundation to be especially important for the transformational journey and is not previously reported in the literature to the best of our knowledge.

Considering the need to modularize the platform ecosystem, we thus propose decision rights partitioning as the overarching capability necessary to seize data-sharing capabilities through external developers. Under unexpected uncertainty (such as addressing the needs of outlaw innovators), this capability can play out through effectual cooperative experiments with developers. When the uncertainty is more recognizable (i.e., how to allow for a more effortless integration with railway data), the logic from the external ecosystem can instead be carried over to the new context.

---

[1] In framing this concept, we took inspiration from the *ecosystem carryover* [40] concept. However, while the original concept concerned utilizing existing ecosystems in new contexts, in our case it was the underlying logics that was replicated.

6.3 Dynamic Capability: Platform Ecosystem Integration

Our final capability concerned how the studied agency could ingest the opportunity offered by external developers into its daily operations. To transform the opportunity presented by external developers, the platform team needed to effectually implement ecosystem requirements (identified during seizing) while drawing on the resources at hand (existing legacy systems). Here, the STA IT management allowed grassroots (the DataCache team) decisions to form on how to go about this complex task. By restructuring existing technology within the administration, the STA was able to inscribe developer needs into the open API platform (subsequently corroborated by the substantial developer adoption and use), a microfoundation we refer to as *autonomous architectural reconciliation*. However, to transform data sharing across the agency, the STA *leveraged internal network effects* through causal execution. While the seizing activities had demonstrated the potential benefits of applying a platform-based data-sharing approach, it needed to be scaled across the administration. Here the platform team utilized the platform's network effects by pushing data curation activities onto integrating teams while allowing an internal user to integrate with DataCache. This way, the existing integration technology platform eventually got increasingly replaced.

In sum, we synthesize these two microfoundations into the transforming capability *platform ecosystem integration*. This capability encompasses the public sector's need to allow for grassroots decisions on how to reconcile developer requirements and existing technology and identify scaling mechanisms that enable data sharing to be propagated throughout the administration.

6.4 Implications for Practice

This study has implications for how organizations can acquire data-sharing capabilities by engaging with external developers in platform ecosystems. First, public organizations need to develop capabilities to sense opportunities for data-sharing in their organizational surroundings, even when it concerns unsolicited use. This includes interacting and collaboratively exploring data-sharing functionality with citizens and private firms through means such as workshops, interviews, or hackathons. Next, to seize opportunities and leverage contingencies public agencies need to efficiently divide tasks between the core platform and data consumers. This involves conducting live, iterative experiments with external developers to refine platform design. The final step in developing external data-sharing capabilities involves having internal teams, who are close to these developers, make key decisions about balancing existing technologies with developers' needs. However, to harness these capabilities internally, organizations must then start sensing inwards and identify opportunities where these platform functionalities could be used to share data within their own organization. Once these opportunities have been identified, the rules for publishing data should be as rigid and scalable as external data, to allow for adaptation and scalability.

7 Conclusions

This paper has explored how organizations acquire data-sharing capabilities by engaging with external developers in platform ecosystems. This inquiry was ignited from an observation that developer practices had subverted a public administration's integration practices. Our analysis revealed that our case organization developed three capabilities and six microfoundations. We label the sensing capability *ecosystem sensemaking* confining the microfoundations platform affordances co-exploration and exploiting data-sharing opportunities. We elicited the seizing *capability decision rights partitioning* encompassing the microfoundations ecosystem alignment experimentation and ecosystem logic carryover. Finally, we identified the transformational capability *platform ecosystem integration* that rested on the microfoundations autonomous architectural reconciliation and leveraging internal network effects.

References

1. Mergel, I.: Digital service teams in government. Gov. Inform. Q. **36**, 101389 (2019)
2. Bason, C.: Leading Public Sector Innovation 2E Co-creating for a Better Society. Policy Press, Chicago (2018)
3. Gil-Garcia, J.R., Helbig, N., Ojo, A.: Being smart: emerging technologies and innovation in the public sector. Gov. Inform. Q. **31**, I1–I8 (2014)
4. Parker, G., Alstyne, M.W.V., Jiang, X.: Platform ecosystems: how developers invert the firm. Manag. Inf. Syst. Q. **41**, 255–266 (2017)
5. Tiwana, A.: Platform Ecosystems: Aligning Architecture, Governance, and Strategy. MK, Amsterdam; Waltham, MA (2014)
6. Wirtz, B.W., Daiser, P.: A meta-analysis of empirical e-government research and its future research implications. Int. Rev. Adm. Sci. **84**, 144–163 (2018)
7. Mergel, I., Edelmann, N., Haug, N.: Defining digital transformation: results from expert interviews. Gov. Inf. Q. **36**, 101385 (2019)
8. Gil-Garcia, J.R., Sayogo, D.S.: Government inter-organizational information sharing initiatives: understanding the main determinants of success. Gov. Inf. Q. **33**, 572–582 (2016)
9. Wang, F.: Understanding the dynamic mechanism of interagency government data sharing. Gov. Inf. Q. **35**, 536–546 (2018)
10. Scholl, H.J. (Jochen), Klischewski, R.: E-government integration and interoperability: framing the research agenda. Int. J. Public Admin. **30**, 889–920 (2007)
11. Eriksson, O., Ågerfalk, P.J.: Speaking things into existence: ontological foundations of identity representation and management. Inf. Syst. J. **32**, 33–60 (2022)
12. Karlsson, F., Frostenson, M., Prenkert, F., Kolkowska, E., Helin, S.: Inter-organisational information sharing in the public sector: a longitudinal case study on the reshaping of success factors. Gov. Inf. Q. **34**, 567–577 (2017)
13. Janssen, M., Estevez, E.: Lean government and platform-based governance—doing more with less. Gov. Inf. Q. **30**, S1–S8 (2013)
14. Baldwin, C.Y., Woodard, C.J.: The architecture of platforms: a unified view. In: Gawer, A. (ed.) Platforms, Markets and Innovation (2009)

146 D. Rudmark and A. Molin

15. Parker, G., Alstyne, M.V., Choudary, S.P.: Platform Revolution: How Networked Markets are Transforming the Economy and How to Make Them Work for You. W.W. Norton & Company, New York (2016)
16. Huang, J., Henfridsson, O., Liu, M.J., Newell, S.: Growing on steroids: rapidly scaling the user base of digital ventures through digital innovation. Mis. Quart. **41**, 301–314 (2017)
17. Tiwana, A.: Platform desertion by app developers. J. Manage. Inf. Syst. **32**, 40–77 (2015)
18. Wareham, J., Fox, P.B., Giner, J.L.C.: Technology ecosystem governance. Organ. Sci. **25**, 1195–1215 (2014)
19. Baldwin, C.Y.: Where do transactions come from? Modularity, transactions, and the boundaries of firms. Ind. Corp. Change. **17**, 155–195 (2007)
20. Foerderer, J., Kude, T., Schuetz, S.W., Heinzl, A.: Knowledge boundaries in enterprise software platform development: antecedents and consequences for platform governance. Inf. Syst. J. **29**, 119–144 (2019)
21. Jussen, I., Schweihoff, J., Dahms, V., Möller, F., Otto, B.: Data sharing fundamentals: characteristics and definition (2023)
22. Teece, D.J., Pisano, G., Shuen, A.: Dynamic capabilities and strategic management. Strategic Manage J. **18**, 509–533 (1997)
23. Hansen, J.R., Ferlie, E.: Applying strategic management theories in public sector organizations: developing a typology. Public Manag. Rev. **18**, 1–19 (2016)
24. Pablo, A.L., Reay, T., Dewald, J.R., Casebeer, A.L.: Identifying, enabling and managing dynamic capabilities in the public sector*. J. Manage. Stud. **44**, 687–708 (2007)
25. Rose, J., Persson, J.S., Heeager, L.T., Irani, Z.: Managing e-Government: value positions and relationships. Inf. Syst. J. **25**, 531–571 (2015)
26. Toll, D., Lindgren, I., Melin, U., Madsen, C.Ø.: Artificial intelligence in Swedish policies: values, benefits, considerations and risks. In: Lindgren, I., et al. (eds.) EGOV 2019. LNCS, vol. 11685, pp. 301–310. Springer, Cham (2019). https://doi.org/10.1007/978-3-030-27325-5_23
27. Barrutia, J.M., Echebarria, C., Aguado-Moralejo, I., Apaolaza-Ibáñez, V., Hartmann, P.: Leading smart city projects: government dynamic capabilities and public value creation. Technol. Forecast. Soc. **179**, 121679 (2022)
28. Teece, D.J.: Explicating dynamic capabilities: the nature and microfoundations of (sustainable) enterprise performance. Strategic. Manage. J. **28**, 1319–1350 (2007)
29. Helfat, C.E., Raubitschek, R.S.: Dynamic and integrative capabilities for profiting from innovation in digital platform-based ecosystems. Res. Policy. **47**, 1391–1399 (2018)
30. Eisenhardt, K.M., Martin, J.A.: Dynamic capabilities: what are they? Strategic Manage. J. **21**, 1105–1121 (2000)
31. Janssen, M., van der Voort, H.: Adaptive governance: towards a stable, accountable and responsive government. Gov. Inf. Q. **33**, 1–5 (2016)
32. Baden-Fuller, C., Teece, D.J.: Market sensing, dynamic capability, and competitive dynamics. Ind. Market Manag. **89**, 105–106 (2019)
33. Mero, J., Haapio, H.: An effectual approach to executing dynamic capabilities under unexpected uncertainty. Ind. Market Manag. **107**, 82–91 (2022)
34. Sarasvathy, S.D.: Causation and effectuation: toward a theoretical shift from economic inevitability to entrepreneurial contingency. Acad. Manage. Rev. **26**, 243–263 (2001)
35. Dubois, A., Gadde, L.-E.: Systematic combining: an abductive approach to case research. J. Bus. Res. **55**, 553–560 (2002)
36. Alvesson, M., Kärreman, D.: Constructing mystery: empirical matters in theory development. Acad. Manage. Rev. **32**, 1265–1281 (2007)
37. Breznitz, D., Ornston, D.: The revolutionary power of peripheral agencies. Comput. Polit. Stud. **46**, 1219–1245 (2013)

38. Daniel, E.M., Wilson, H.N.: The role of dynamic capabilities in e-business transformation. Eur. J. Inf. Syst. **12**, 282–296 (2003)
39. Parker, G., Alstyne, M.W.V., Jiang, X.: Platform ecosystems: how developers invert the firm. SSRN Electron. J. **41**, 255–266 (2016)
40. Adner, R.: The Wide Lens: A New Strategy for Innovation. Portfolio/Penguin, New York (2012)

Digital Sovereignty

Shaping a Data Commoning Polity: Prospects and Challenges of a European Digital Sovereignty

Stefano Calzati[✉] [iD]

Delft University of Technology, 2626BZ Delft, The Netherlands
s.calzati@tudelft.nl

Abstract. The concept of "digital sovereignty" has gained momentum due to the emergence of a multipolar geopolitical scenario based upon different visions of today's digital society. In this scenario, the United States, China, and the European Union are major players, each pursuing their understanding of digital sovereignty and their approach to digital transformation. The EU conceives of digital sovereignty as technological autonomy from other competitors, and to achieve this it has carved for itself the role of international regulator. De facto, however, the EU enacts an individual-centric and economic-driven digital strategy that hinders the possibility of a fully-fledged European digital sovereignty. Notably, the concept fails to embed the collective-level dimension proper to sovereignty as such. To tackle this, the paper explores data commoning as the basis for shaping a well-formed European polity, key to its digital sovereignty.

Keywords: digital sovereignty · European Union · data commons

1 Introduction

In contemporary political theory, the concept of sovereignty is one of the most debated, being regarded as a contested concept that, at once, has undergone deep changes in meaning – to the point of eroding its epistemological relevance (Agnew 1999) – and one which remains a pivotal feature of the contemporary geopolitical landscape (Werner & De Wilde 2001). An operational definition for the present paper considers sovereignty as "a process in which a group of people within a defined territory is moulded into an orderly cohesion through the establishment of a governing authority that can be differentiated from society and which is able to exercise an absolute political power" (Loughlin 2018). This definition is useful – as we shall see – for two reasons: on the one hand, it identifies the key attributes of sovereignty, notably: territory, authority/power, and community; on the other hand, by departing from a normative (Westphalian) understanding of the term, it stresses the procedural nature of sovereignty as an emergent feature of/within any polity. Based on this premise, the paper explores 1) what does the European Union mean

N. Edelmann et al. (Eds.): ePart 2023, LNCS 14153, pp. 151–166, 2023.
https://doi.org/10.1007/978-3-031-41617-0_10

with sovereignty in the context of digital transformation[1]? 2) How does the concept of a sovereign digital transformation impact on the achievement of the EU's digital strategy?

In tackling the first question, the paper follows up on the "linguistic turn" in political theory, whereby concepts once considered as nominally self-sufficient, have been unveiled to be socially constructed, thus depending on pragmatic use for their own validation. "Meaning", then, stands here not much for a dictionary-like definition, but for how the concept is adopted in official documents – EU's policy documents, directives, and regulations – and with which consequences. Concerning the second question, the goal is to investigate the extent to which the meaning of digital sovereignty adopted by the EU hinders and/or favors the pursuing of its digital agenda, based on a strategy that seeks a balance between individual fundamental rights and economic competitiveness, social inclusiveness, and environmental sustainability (von der Leyen 2020).

To do so, a critical review is conducted of latest policy-orienting documents and pieces of legislation published by the EU as part of its digital strategy. According to Grant and Booth (2009), a critical review is a method that delivers "analysis and conceptual innovation" for future informed research and practice. Hence, the present critical review is not exhaustive in scope, but rather identifies (discursive) patterns which then establish, de facto, the way to follow when it comes to governing the digital transformation within the EU. Since the analysis will highlight a discrepancy between the EU's digital strategy and digital sovereignty, the paper will advance some suggestions on how the EU might purse a more coherent sovereign digital strategy. As a note, the paper is conceptual in nature and further testing of its key tenets is needed.

The paper is divided in five sections: Sect. 2 provides an overview on sovereignty as a contested concept in political theory; Sect. 3 explores the concept of digital sovereignty with a focus on the EU in relation to its main technological competitors; Sect. 4 is concerned with how this conceptualization of digital sovereignty relates to the EU's digital strategy, highlighting a potential misalignment; Sect. 5 outlines key conceptual tenets to enact a fully-fledge European digital sovereignty; Sect. 6 summaries the key points and points towards further research.

2 Sovereignty

Based on a normative Westphalian understanding, the nation-state has been traditionally the privileged locus of sovereignty. In fact, in the nation-state we witness the overlap of the triad territory, authority/power, and community, which is at the core of a legal and political characterization of sovereignty. However, especially since the second half of the 20th century, the endowment of the nation-state with such sovereign legitimacy has gotten increasingly contested along at least four axes: 1) the misalignment between sovereignty *de iure* and *de facto*; 2) the extension of sovereignty to supra- and sub-national dimensions; 3) the clash between the state and private actors; 4) the consolidation of an infrastructural global network cutting physical-virtual borders. While these axes are deeply intertwined, they are disentangled here for analytical purposes.

---

[1] Here "digital transformation" refers to the sociotechnical effects of digitalization, intended as the translation of physical reality into a set of 0s and 1s, through data-driven technologies.

2.1 Sovereignty *de iure* and *de facto*

The misalignment between sovereignty *de iure* and *de facto* is double-faced. On the one hand, what has come to be regarded as nation-state sovereignty *de facto* since the 1648 Peace of Westphalia does not find a coherent reflection *de iure*. Beaulac (2004) notes, indeed, that those institutional, legal, and political arrangements usually associated with the Westphalian order do not find a statutory reflection in the treaties of Münster and Osnabrück which are regarded as the founding documents of that same order. In fact, these documents address issues of religious tolerance, territorial settlements, and legal powers; and yet, they have become *de facto* the pillars for the crystallization of nation-state empires.

On the other hand, sovereignty as established *de iure* has been (and still is) repeatedly redesigned by the realpolitik of global geopolitics. Suffice to think, in this regard, to centuries of colonization, unilateral borders transgression (the latest case is the war in Ukraine) and, more broadly, the enactment of (il)legitimate state power beyond the limits established by law and territory. In this respect, Krasner (1999) arrives to characterize sovereignty as "organized hypocrisy" to the extent that what is declared in official documents gets repeatedly contested, if not subverted, by practice.

2.2 Sovereignty Across Scales

Concerning the second axis, economic trades on a global scale, alongside the reshaping of state's functions, have been identified as factors that diffused the coalescence of territory, authority/power, and community into the nation-state. In recent decades, a new galaxy of authorities and communities have contested the nation-state as the sole legitimate beholder of sovereignty. On the one hand, Ilgen (2003) contends that under the pressure of economic globalization central governments have forfeited part of their power and authority to external actors, leading to forms of multilayered governance, at both supra- and sub-national levels. Current examples of this in Europe are the increasing autonomy allocated to municipalities and the establishment of the EU as a supranational framework. The problem, in this latter respect, is that the EU has designed for itself the role of international regulator without however sufficiently binding power: as Soare (2023) notes, a "perpetual gap" endures between "EU policymaking and national implementation and compliance." In other words, the EU acts as a legislator without full legitimacy, due to a discrepancy between its authority and a well-formed polity upon which to exert power. Irion (2023) adds that, in the context of digital transformation, "the EU is currently producing many new pieces of legislation on digital issues, which may overstretch the capacity of proper implementation by stakeholders and enforcement", thus leaving behind a patchworked policy landscape hard to harmonize.

At the same time, the push towards a diffusion of sovereignty across scales has been favored by new managerialist approaches to the public sector (Pollitt et al. 2007). With the promise of performing more efficiently towards citizens (mostly regarded as consumers), the state apparatus has embraced economic-driven approaches to the delivery of public services, ultimately dislocating functions that were once its own prerogative to sub-national bodies, external organizations, as well as private actors. This links more closely to the third axis of contestation of nation-state sovereignty.

2.3 Transnational Sovereignty

In recent decades, global trade has become an increasingly entangled affairs manifested by the emergence of transnational relations that cut across and remold nation-states' sovereign legitimacy. As Wen (2021) writes, "the development of the global economy has been characterised by the transition towards transnationalized capitalism, within which information and communications technologies have increasingly played a pivotal role in restructuring the global capitalist system." Hence, an accurate understanding of such scenario through the lens of sovereignty, and past the cornerstone of the nation-state, requires undoing conceptual dichotomies such as global-local, as well as public-private.

In this respect, Wasserman (2018) observes that at stake is the remaking of global power relations that "have prompted different ways of thinking about categories such as the 'South,' the 'global,' the 'local'." For instance, the "going out" phase of Chinese companies across the globe – after years of internal state-supported consolidation – has been perceived, especially in the West, as a form of soft power, if not colonisation *stricto sensu*. From this perspective, Chinese companies have been depicted as the *longa manus* of the government in different regions of the globe. However, ground evidence shows a more fine-grained scenario where Chinese companies' trade does not follow one unique party line, but it is rather the result of different and sometimes conflicting visions. Xu (2014) writes that "Chinese firms have created serious challenges for the Chinese government to regulate them at home and overseas"; while Gu and colleagues (2016) reveal the "proliferation of Chinese businesses acting independently or, depending upon ownership, semi-independently of the Chinese state."

To emerge then are federated forms of globalization – contested internally as much as externally – in which the circulation of goods – including data – and subjects, as well as the adoption of regulations, depend on the entrenchment of multifactorial trends building on competing agendas, authorities, powers, and territories. This demands to approach sovereignty by keeping into account the entangled nature of such federated forms of globalisation (Calzati 2020).

2.4 ICT-Based Sovereignty

As seen with transitional sovereignty, ICTs have become pivotal in redesigning sovereignty across public and private actors. ICTs are, at once, cause and effect of the (re)wiring of the globe under commercial pushes. Concretely, the framing of ICTs within a sovereign geopolitical perspective, brings with itself forms of power asymmetry. Studies have shown the "misalignment" between the Internet as a commons infrastructure and the legitimacy of sovereign powers (Mueller 2019). Traditional categories such as "market" and "state", "national" and "international", may no longer be sufficient to account for today's tech-based geopolitics.

For instance, Yu and Goodnight (2020) argue, with specific regard to China, that: "China's so-called Intranet also reveals entanglements with foreign capital, foreign technology, foreign markets, and foreign labor." More specifically, while the US have been heralded as defending a multi-stakeholders "free Internet" and China, on the contrary, as advancing a multilateral "sovereign Internet", the scenario is more complex, with these two competitors often finding agreements behind the scenes especially when it comes to the very basic principle of surveillance through data. As an example, Gagliardone (2016) writes that in 2012 "representatives at the WTSA were swift in passing a new standard on the 'Requirements for Deep Packet Inspection in Next Generation Networks,' or 'Y.2770.' Discussions happened behind closed doors and no drafts were circulated before a final decision was made, attracting criticism on the lack of transparency." In other words, the agreement on the standardization of deep packet inspection (DPI) did find wide consensus among different actors, even between those (supposedly) heralding opposing views on the Internet's governance.

This means that ICTs have impacted on nation-state sovereignty in multifarious ways, reshaping established power relations, fostering new alliances based on contingent interests, as well as creating preconditions for new asymmetries within and beyond the nation-state.

3 Digital Sovereignty

This brings us to explore the concept of "digital sovereignty". According to Kushwaha and colleagues (2020), "digital sovereignty" is a concept that is mid-way between the broad idea of "technological sovereignty" and the narrow idea of "data sovereignty". In fact, Irion (2023) notes that digital sovereignty remains "conceptually fuzzy and is used to animate a wide spectrum of geopolitical, normative and industrial ambitions"; while Timmers (2023) identifies a stack of at least six layers – "key enabling technologies", "semiconductors", "networks", "data", "cloud services" "apps" – to which digital sovereignty might apply.

Beyond these analytical distinctions, at the heart of the matter is control over data and tech infrastructures (Hummel et al., 2021). Overall, apart from those countries able to chart their own course, the risk of being co-opted by major global tech actors, private or public, is high. At stake, then, is an issue of governance and, most notably, how to account for the distribution of data/tech power across a diversity of actors (Micheli et al. 2020). As scholars (Glasze et al. 2022) point out, technology governance becomes part and parcel of geopolitics when power relations heavily influence how a technology is developed, implemented, controlled, and used.

From this perspective, digital sovereignty can be best regarded as a macro entangled cyber-geopolitical dimension which contests and resists linear (agent-structure) readings on which nation-state sovereignty rests. It is a whole *ecosystemic procedural* dimension that comes into being (see also Sect. 5). For instance, while US corporations tend to dominate Internet services and software, the "ownership" of hardware infrastructures depends on an imbrication of actors. As an example, the transpacific FASTER cable system, connecting the US and several cities in Japan, China, and Korea, was jointly developed by Chinese, American and South Asian private companies. It is evident that such diverse composition questions the epistemological robustness of sovereignty to its state-centric (Westphalian) roots, demanding to account for the enmeshment between technology and geopolitics that digital sovereignty entails.

Luciano Floridi (2017) calls "cut-and-paste" the logic at the basis of digital transformation: "the digital cuts and pastes reality in the sense that it couples, decouples, recouples features of the world". This has deep repercussions on sovereignty as traditionally conceptualized in terms of territory, authority/power, and community.

To begin with, data-driven technologies frame the subject – and turn it into a data subject – regardless of its location: access to a digital infrastructure is all that is needed and sometimes this might be independent from the subject's own will. This implies a schism between the subject's physical and virtual existence, which goes in hand with a redefinition of their rights. The emergence of e-residency programs developed by European countries (e.g., Estonia and Portugal), especially on the wave of the pandemic, is a case in point. E-residents are bestowed with a (location-independent) digital citizenship attached to the country they apply to. For becoming e-residents the monadic fusion of presence and territory is no longer required, insofar as any subject can potentially apply to e-residency programs from/to anywhere in the world. Similarly, the legislation of the country to which the subject has applied – including supra-national frameworks – comes to extend beyond its physical territoriality and it does so by enforcing a digitization of both the (data) subject and physical placedness, which gets virtualized into a non-local space. More broadly, this is a good example of two simultaneous "decouplings", as Floridi (2017) calls them, made possible by the digital transformation: that between location and presence, on the one hand, and that between law and territoriality, on the other. In fact, e-residency programs emerge out of the splitting of these binomials and by leveraging on their recombination, producing an entanglement of its own between the data subject and a set of actors – banks, public authorities, as well as other e-residents – with which the data subject inevitably gets enmeshed.

"Digital sovereignty", then, is a multifaceted concept that cannot be reduced to a linear mapping of the actors involved and their relations. This is so because it is the fundamental attributes of sovereignty – territory, authority/power, community – that the "digital" contributes to remix. It is only when the notion of digital sovereignty is contextualized and approached as an ongoing process needing finetuning that it becomes possible to question who claims power, on whom such power is exerted, for what purposes, and with which consequences.

3.1 Digital Sovereignty of/in the European Union

In February 2020, Ursula von der Leyen (2020) defined digital sovereignty as the capability "to make its own choices, based on its own values, respecting its own rules' in the field of tech." This is a definition of sovereignty that puts the stress on autonomy in the sense of technological self-determination. In fact, already in 2016, the Council of the European Union defined strategic autonomy, which is also applied to digital sovereignty, as the "capacity to act autonomously when and where necessary and with partners wherever possible." These two claims are relevant for two reasons.

Beginning with the latter claim, partnerships are considered as key enablers toward strategic autonomy; and yet, when this understanding is applied in the context of digital sovereignty, partnerships remain more of a wish than a concrete strategy: as Soare (2023) explains unambiguously "the EU does not have a clear idea of what a new approach to tech partnerships should look like [and] it lacks a balanced approach to partnerships." So far, the EU has adopted a rather passive approach in the shaping of its digital sovereignty, largely interpreting "autonomy" as lack of interference from foreign actors. This is mostly done against the disproportionate data grabbing of US-American tech companies, as well as the deployment of infrastructural networks by

Chinese ones. On the one hand, companies such as Microsoft, Google, Amazon have opened datacenters around the continent, to the point that the EU (2019) has warned against the "digital dependency on non-European providers and the lack of a well-performing cloud infrastructure respecting European norms and values", which can de facto be considered as a form of colonization. On the other hand, while initially European countries welcomed Chinese giant Huawei to roll out its 5G network across the continent – partly to reduce dependency on the US, partly due to the unclear national and supra-national legal overlaps (Robles-Carrillo, 2023) – later the project was halted due to possible security risks at national and supra-national levels.

Recent pieces of legislation such as the Digital Service Act (DSA) and the Digital Markets Act (DMA) represent two steps in the direction of a more robust and binding regulation of large private tech platforms and data service providers with the goal to both create a safer digital space in which the fundamental rights of all users are safeguarded, as well as to establish fairer rules towards the boosting of innovation and competitiveness. Other initiatives that aims to promote the EU's autonomy in matter of digital transformation are the "5G toolbox", that is, a comprehensive European framework designed through the coordination of Member States' national policies in matter of adoption and deployment of 5G network; and the European Chips Act aimed to unburden the EU from its dependence on foreign actors as far as the supply of advanced semiconductors is concerned (although this, in turn, questions the impact of such Act on the sovereignty of a third country – Taiwan – which is one of the biggest suppliers of semiconductors globally and a country whose sovereignty is threatened by China).

Leaving aside the difficulty of effectively enforcing all these pieces of legislation in a harmonious way, these initiatives show limitations in two respects. On the one hand, the interpretation of digital sovereignty in terms of lack of interference from foreign actors is not sufficient for achieving a fully formed European digital sovereignty. In this regard, the EU needs to adopt a more proactive stance for instance by promoting bilateral cooperation to defend its own digital assets and develop symmetric power relations, especially with the US and China.

A case in point is the redefinition of the EU open data policy towards non-EU actors, either private or public. China's nation-state approach towards the regulation of its own data landscape is well-known by now; what is less know, however, is that China's initiatives on this matter – recently, the Personal Information Protection Law (PIPL) and the Data Security Law (DSL) – have two complementary effects: to expand the (extraterritorial) outreach of its own legislation, as well as disempower foreign measures negatively impacting its own interests. It is a digital sovereignty that entangles privacy issues with national security; moreover, it is a sovereignty that is centrifugal as far as the outwards impact of China's legislation is concerned, and centripetal as far as the inward securization of its assets – data, infrastructures, services – is concerned.

This is just the latest instantiation of a series of tactical decisions involving the US and China. Already in 2018, the US Cloud Act, which was passed after the adoption by China of its Cybersecurity Law in 2017, represents a policy move that, under the guise of data localization and protection, has an eminently transnational character; to which China responded with its PIPL and DSA (both approved in 2021). The point is that such decisions might also have undesirable commercial repercussions, as it was the case with the Trump's administration issuing a commercial ban against Huawei in 2019 through a security order. While such ban did impact Huawei's economic performances, some US companies found themselves in an odd position as both Huawei's commercial partners and actors bound to national "duties". This eventually led some US companies to bypass the ban, with FedEx suing the US Commerce Department and Google warning the Trump administration that the ban would constitute a national security risk in its own right. This shows the extent to which "states will increasingly face difficult policy decisions with regard to deciding how best to balance competing sovereign interests" (Kushwaha et al., 2021). In this scenario the EU is reactive rather than proactive, mostly responding to the consolidation of Chinese market dominance and the conjoint effort by the US to renationalize supply chains (Broeders & Kumiska, 2023). Hence, the highlighted lack in the EU of a clear strategy of external cooperation is a clear limitation and might constitute a testbed for designing more symmetric agreements with foreign partners based on shared values and equipollent legislations. These agreements could leverage on the EU's strength as an international regulator and on its digital assets, such as the open data policy which currently risks asymmetrically benefitting a diverse array of actors without sufficient return and safeguards. On this, Voss and Pernot-Leplay (2023) contend that an adequacy determination between China and the EU in matter of data transfer is currently not possible, meaning that a power asymmetry endures.

On the other hand, the EU's policy initiatives tend to privilege the preservation of individual rights, such as privacy – e.g., the DMA aims to "create a safer digital space in which the fundamental rights of all users are protected" – over collective-level rights – such as democratic participation – as well as the pursuing of economic competitiveness – e.g., the DMA aims to "establish to establish a level playing field to foster innovation, growth, and competitiveness" – over the creation of social value. To better understand why such double focus is limiting, von der Leyen's definition of digital sovereignty provided above comes in handy: "to make its own choices, based on its own *values* [italics added]." What are these values? In presenting Europe's digital future in 2020, von der Leyen spoke of the need to enact a digital strategy in which technology works for people, promotes a fair and competitive economy, and foster an inclusive and sustainable society. As we will see below, currently "people" have been reduced to the individual/consumer and central stage has been taken by economic interests and actors, while social and collective-level concerns have fallen into the background.

4   The EU's Strategy and the Gap with Digital Sovereignty

The EU has adopted a human rights-based approach (HRBA) to governing digital transformation (Brown, 2019), which, while being pivotal for preserving individuals' integrity before digital transformation, especially in terms of freedoms and privacy, risks systematically overlooking societal and collective-level values – such as social inclusion, environmental sustainability and digital sovereignty – which cannot be boiled down to individuals and their rights (Smuha, 2021; Taylor et al., 2017; Viljoen, 2021). For instance, Taylor and colleagues (2017) discuss the idea of "group privacy" and the need to redesign current legal frameworks, starting from the acknowledgement that data-driven technologies address and impinge on groups-as-collectives besides and beyond individuals. Going further, Viljoen (2021) notes that the individualistic vision behind the current EU approach does not account for the relational nature of data and the consequent trade-off effects that data re-use involving two subjects might have on unaware third parties. On this wave, Smuha (2021) suggests taking inspiration from environmental law for tackling potential collective-level effects caused by digital transformation, such as the erosion of the legitimacy and functioning of the rule of law, which can be neither accounted for nor mitigated by the current EU approach to digital transformation Hence, to the extent to which HRBA does constitute the fundamental baseline to citizens' autonomy, it might be insufficient to enact a fully-fledged digital sovereignty, requiring to shape a well-formed polity to be legitimate.

A partial response comes from the Declaration on Digital Rights and Principles of the EU, which defends "a European way for the digital transition, putting people at the center." Notably, what the DDRP does is to pin down six principles – 1) preserve people's rights; 2) support solidarity and inclusion; 3) ensure freedom of choice; 4) foster democratic participation; 5) increase safety, security, and empowerment of individuals; 6) promote sustainability – which equally split between a half (1, 3, 5) focusing on the individual and the other half (2, 4, 6) pertaining to society as a whole. Hence the DDRP does strive to strike a balance between subject-centric and collective-centric dimensions (Calzati, 2022). However, so far, such balance has not been operationalized, constituting a critical point of contention for the enactment of a European digital sovereignty.

Recent documents have laid the ground for the establishment of an EU digital single market (DSM), as the arena where the digital strategy will play out. In this sense, the 2021 Digital Europe Programme speaks of "the importance of building a thriving ecosystem of private actors to generate economic and societal value from data, while preserving high privacy, security, safety and ethical standards." This statement is significant because it clearly places private actors at the centre of the market, endowed with the task of creating economic and societal value. On this point, Taylor (2021) warns against the notorious difficulty of "establishing meaningful accountability for the private sector" which hinders an effective public scrutiny of how tech companies operate, for which purposes, and with which results. The risk is to see the conflation between public value created by the public sector and public value created by businesses "despite the profit interests involved and the different regulatory architectures occupied by firms and government" (Taylor, 2021). While it might occur that private companies do deliver public value, this can hardly occur on a systemic basis, that is, one that keeps into account collective-level tradeoffs beyond a cost-benefit logic.

The subsequent 2022 Digital Europe Programme provides a clearer characterization of the emerging DSM. Here the European Commission speaks of "the deployment of (…) common data spaces, based on (…) a data infrastructure with tailored governance mechanisms that will enable secure and cross-border access to key datasets in the targeted thematic areas." The DSM is a secured technical backbone revolving around private actors and achieving economic-driven and GDPR-compliant data sharing. In a similar vein, in 2019 the GAIA-X project was launched by a nonprofit foundation with the goal to "enable a sovereign decision on data-based business models" and to promote "common models and rules for data monetization", as well as "cross-industry cooperation to create federal, interoperable services" (BMWi, 2020). Most importantly, behind

GAIA-X is a consortium founded by 22 German and French companies supervised by the German Federal Ministry for Economic Affairs and Energy. Albeit being a no-profit foundation, GAIA-X's governance raises concerns about the way in which sovereignty can be actually guaranteed as a collective principle, since it gets dislocated to private actors and placed in the hands of only two countries, without dutiful consultation and orchestration. Literature shows (Monti, 2023) that tech-centered and market-driven policies are relatively weak tools for pursuing digital sovereignty and strategic autonomy, in that these policies can hardly become compasses for political action.

Overall, the individual-centric and economic-driven approach of the EU to digital transformation is at odd with a full-fledged idea of digital sovereignty that maintains a societal and collective outlook by default, able to cut across scales and involving diverse actors. At stake is the need to design an arena that moves away from prioritizing either certain actors – private or state actors – or values – oftentimes economic competitiveness over social inclusiveness or environmental sustainability – to rather enact a systemically balanced ecosystem.

5 An Ecosystemic Proposition

An ecosystem is characterized by homeostasis, that is, the balanced interaction between biotic and non-biotic elements within an environment. This implies that the ecosystem's behavior cannot be studied by isolating either elements or interactions; rather, it must be studied in its entirety. While largely related to the natural world, the notion of ecosystem has also been applied to other settings, such as the digital landscape (van Loenen et al., 2021). To endorse an ecosystemic vision towards the governing of digital transformation means to seek a fair governance in which all actors' interests are accounted for and negotiated. In other words, fairness underscores here the systemic trading off among different interests in view of an overall equilibrium.

This understanding of fairness overcomes both a reductionist and an essentialist definition of the term. Within the first group fall those attempts which seek to provide a mathematical definition of fairness, overlooking its contextual dependency. On the other hand, an essentialist standpoint does account for the context-dependency of fairness, and yet it still considers it as a core quality of a given technology or data process, failing to produce a comprehensive enactment of fairness within a given scenario.

To shift towards an ecosystemic understanding of fairness it is worth looking at how the EU defines this term in the context of the development and implementation of data-driven technologies. Notably, the European Commission disentangles fairness as both a substantive and procedural affair. On this point, Rochel (2021) notes that as a structuring principle of the GDPR "fairness" is "linked to principles such as proportionality and other procedural dimensions of a balancing exercise involving rights and interests." This highlights well the fact that, beyond the matching of certain requirements, fairness is an act of balance based upon the recognition and negotiation among different interests and rights on a flexible and rolling basis. Hence, a governance framework that aims to regulate a data ecosystem fairly identifies roles and rules to represent the data interests of all actors, as well as mechanisms to adjudicate situations where conflicts among actors and/or values might arise. Most importantly, such ecosystem shall be regarded not much as an arena where different players are connected, but as a process that constantly reshapes its own power relations. It is in this respect that the definition of sovereignty provided in the introduction is particularly fitting in that it regards sovereignty as a process, more than a state of affair. How to enhance such process will occupy the remnant of the paper.

5.1 Data Commoning: Conceptual Tenets for a European Digital Sovereignty

To tackle the situation, it is necessary to rethink democratic participation *through* and *about* digital transformation. Scholars (Zygmuntowski, et al., 2021) have hinted at the promise of designing an EU data governance that is based on the logics of the commons.

Originally, the commons referred to natural resources characterized by non-excludability (i.e., difficulty or impossibility of forbidding access and use of resources to any potential beneficiary) and rivalry (i.e., the use of resources depletes them and reduces further use by others). Ostrom (1990) showed that the self-management of resources by communities can be more effective than market-driven or state-led approaches, provided that principles and roles are designed and abided to. Moving towards the "second wave", by now the commons has been applied to non-natural resources, such as data (Dulong de Rosnay & Stalder, 2020). Today, Data Commons (DC) initiatives aim to counteract and/or repurpose the centralized ownership and use of data – either by tech companies or states – by giving these back to citizens, with the goal to foster sustainable collective data practices (Morozov & Bria, 2018). Overall, DC defines a self-organizational management of data and infrastructures which is non-appropriative by default (knowledge, assets, and outputs are not owned, in the commercial sense of the term, but summoned up and recirculated); collaborative by design (it considers all actors and links within the ecosystem as integral and necessary to the system's flourishing), and collectively sustainable in its goals (indeed, common goods for the community) (Calzati, 2022). This means that the creation of social value – in either tangible or intangible forms – is regarded as desirable on an equal footage with economic value, which is then recirculated within the system.

Bloom and colleagues (2021) suggest how Ostrom's design principles for managing natural resources might be transposed in the context of data initiatives, in terms of access, management and adjudication of resources. However, their standpoint remains anchored to a normative understanding of data as a resource, preventing an effective tackling of data through the lens of the commons (Sanfilippo & Frischmann, 2023).

More useful is to move beyond the conception of the commons as a resource – a thing – to accommodate the idea of "commoning" (de Angelis, 2017) as a sociotechnical process. As de Angelis (2017) notes "commons are not just resources held in common, or commonwealth, but social systems [of] ongoing interactions, phases of decision making and communal labor process." The shift is crucial when applied to data. Indeed, differently from natural resources, data do not pre-exist in nature. Instead, data are a fully artificial (human and/or tech-created) construct that exists in the very moment in which a certain (sociotechnical) process is enacted. Hence, data-as-resource are unique in that they manifest an entangled nature: if one stresses the informational constituency of data, then data are a virtual entity and are potentially distributable globally; if one stresses the technical constituency of data (from collection to storing and use), then data are material entities whose allocation and circulation can be favored or hindered in many ways. The hybrid nature of data is also responsible for tensions at legal level: someone can claim ownership over data even without control (and vice versa), stressing either the informational (e.g., European doctrine) or technical constituency (e.g., US doctrine) of data. The commoning of data, then, requires a paradigmatic shift in the way to think and manage data: data are always created under certain (sociotechnical) conditions, used for certain purposes, in certain contexts, by certain actors, and with certain results. Decisive, in this regard, is the boundary of the data commoning, and how this boundary negotiates the hybrid nature of data.

In other words, at the core of data commoning is a *certain* idea of polity. A (data) polity is a fractal concept as far as its scale is concerned in that it depends on the interplay among three components: infrastructures, institutions, and people. As long as these components are ideally co-extensive (i.e., they overlap), then authority and territoriality are fully legitimate, as the exercise of power coincides with (and can be scrutinized in) the interest of the whole community. Whenever the co-extensiveness of the three is not guaranteed, as it is often the case – e.g., an international actor comes in play in a given data polity under international market laws – then we have a weakening of legitimacy because of a discrepancy between authority and territoriality. This inevitably implies that the blossoming of a given polity is subjected to ongoing (re)negotiation. Already today, local, national, and supra-national legal frameworks are in place for disentangling individual and collective interests concerning the access and (re)use of (personal) data. This is so because "general interest" is an entangled concept: from an empirical perspective, the concept reflects the diversity of interests of all actors involved in a given situation; from an ethical perspective, it constitutes the synthesis (not necessarily the sum) of all actors' interests. In fact, such synthesis is never given once and for all; rather, it is based on ever-changing discontinuities across the polity and among its actors. Concretely, this demands the design of an iterative process able to reflect upon itself – and its own condition of existence – in a democratic way. The term *communitas* etymologically identifies, not much the sharing of "things", but a *duty to come together* (*cum + munus*). This suggests that a data polity moves across a spectrum that fairly negotiates and/or adjudicates between *public* and *private actors*, *collective* and *individual* rights and values, as well as *informational* and *technical* constituencies of data. Most importantly, a data polity comes with rights and responsibilities for contributing to and maintaining the pooling of data; it is as much an issue of *control* as of *care*: in fact, the balancing act between these two poles is what might define an indigenously European digital sovereignty.

For the present discussion the outer horizon of the EU's data polity is European citizens, institutions, and territory. As seen, such characterization shall not be considered monolithically, but as a dimension in constant articulation across scales and contexts. At the same time, the very fact of identifying a European polity turns the issue of digital sovereignty on its head, starting from the premise that there exists such a polity, and it has a continental outreach: as Broeders and Kaminska (2023) argue, "member States must realise that policy coordinated under the EU umbrella is more fruitful economically and geopolitically than national actions." Concerning what lies beyond the EU, digital sovereignty shall be based on systemic fairness as discussed above, notably by drafting symmetric agreements as far as the (societal and economic) value of data is concerned, as well as based on equipollent legislations as far as the protection of fundamental rights is concerned. This leads to suggests that, based on the categorization of the United Nations Conference on Trade and Development, internally the EU could promote a "light-touch" or fully open data sharing, while externally it should articulate a spectrum going from "strict localization" of data to "conditional soft transfer" based on symmetric agreements. To do so, however, the EU needs to establish for itself a well-formed polity. Three axes are at stake: 1) which actors; 2) which interests; 3) which features of data.

The term "public actors", on the one hand, cuts across scales – from sub-national to EU levels – aligning to the discussion on sovereignty discussed above; on the other hand, it involves both institutional *and* non-institutional actors. In fact, a heterogeneous galaxy of actors does contribute to inform data commoning: NGOs, non-profit organizations, data intermediaries, data stewards, etc. (including free riders). This heterogeneous galaxy is increasingly acknowledged – yet, not operationalized – by the EU (e.g., in the Data Governance Act), for instance identifying data altruism organizations and data cooperatives. The term "private actors" refers to small and medium enterprises, as well as big tech giants and public undertakings. In this respect the modulation of the commoning might depend on several factors, such as size, position in the market, tasks, revenues, etc. On this point, the commoning can builds upon the criteria identified by the DMA and DSA as a compass; however, to reach an effective identification and operationalization of all these actors and factors requires further research.

Concerning individual and collective dimensions, processes of arbitration shall be designed to disentangle and/or adjudicate the most fitting commoning approach whenever conflicts between interests and values arise. Since the Open Data directive, the EU has acknowledged that "means of redress should include the possibility of review of negative decisions." More recently, the Data Act speaks of "settlement bodies" to ensure "alternative ways of resolving domestic and cross-border disputes in connection with making data available." Yet, how to properly design such bodies so that they harmonize legal frameworks across scales and in different contexts remains an open issue.

Concerning the informational and technical constituency of data, governance mechanisms must be designed to either negotiate between the two constituencies of data or disentangle and give priority to either one of the two. The commoning modulation, then, impacts on different levels of access and management (including reuse) of data, depending on the kind of initiative at stake, the type of data, and the actors involved, and their goals. In this sense, the spectrum of commons licenses shall be regarded as a starting point towards all-exhaustive framework able to finetune to different scenarios.

6 Conclusion

The paper firstly discussed how the emergence of a global networked society has shaped digital sovereignty into a cyber-geopolitical entangled affair. Secondly, the paper delved into the EU's understanding of digital sovereignty as technological autonomy and linked this to the EU's digital strategy. On the one hand, such understanding is rather passive compared to competitors and ultimately insufficient to protect and enhance the EU's digital assets (RQ1); on the other hand, by pursuing an individual-based economic-driven digital strategy, the EU hinders the possibility of achieving a fully-fledged digital sovereignty, which maintains a collective horizon by definition (RQ2).

To counteract this, the paper builds upon the idea of data commoning, as a sociotechnical process that can favor the consolidation of a European data polity. Notably, data commoning can accommodate a fair representation, negotiation, and, if needed, adjudication of individual data interests, while keeping a societal and collective outlook. Key, in this regard, are 1) the involvement of both institutional and non-institutional actors on a rolling basis; 2) the definition of data arbitration processes able to cut across scales and contexts and negotiate or adjudicate between individual and collective dimensions; 3) the identification of access rights and managing responsibilities modulated on the premise of pooled data as both informational and technical constituencies. Recent EU's pieces of legislation begin to address points 1 and 2; yet, *how* to systemically design such involvement and arbitration are open questions. Point 3, instead, remains uncharted and requires further investigation.

While laying down the foundation of a European data polity and a possible way to enact it beyond market- or state-oriented approaches, this paper is conceptual in nature. As such, the identified coordinates enabling the envisioned data polity need to be operationalized, ideally in living lab scenarios or through action research, to be validated.

References

Agnew, J.: Mapping political power beyond state boundaries: territory, identity, and movement in world politics. Millennium **28**(3), 499–521 (1999)

Werner, W.G., De Wilde, J.H.: The endurance of sovereignty. Eur. J. Int. Rel. **7**(3), 283–313 (2001)

Loughlin, M.: Ten tenets of sovereignty. In: Walker, N. (ed.) Relocating Sovereignty, pp. 79–110. Routledge, New York (2018)

von der Leyen, U.: A union that strives for more: My agenda for Europe (2020). https://ec.europa.eu/info/sites/default/files/political-guidelines-next-commission_en_0.pdf

Grant, M.J., Booth, A.: A typology of reviews: an analysis of 14 review types and associated methodologies. Health Info. Libr. J. **26**(2), 91–108 (2009)

Beaulac, S.: The power of language in the making of international law: the word sovereignty in Bodin and Vattel and the myth of Westphalia. Brill (2004)

Krasner, S.D.: Sovereignty. In: Krasner, S.D. (ed.) Sovereignty. Princeton University Press, Princeton (1999)

Ilgen, T.L. (ed.): Reconfigured Sovereignty: Multi-Layered Governance in the Global Age. Ashgate Pub Limited, Aldershot (2003)

Soare, S.: How to achieve digital sovereignty – a European guide. In: Digital Sovereignty: From Narrative to Policy? pp. 19–24 (2023). https://eucyberdirect.eu/research/digital-sovereignty-narrative-policy

Irion, K.: The general data protection regulation though the lens of digital sovereignty. In: Digital sovereignty: From narrative to policy? pp. 53–57 (2023). https://eucyberdirect.eu/research/digital-sovereignty-narrative-policy

Pollitt, C., Van Thiel, S., Homburg, V. (eds.): New Public Management in Europe. Palgrave Macmillan, Basingstoke (2007)

Wen, Y.: The Huawei Model: The Rise of China's Technology Giant. University of Illinois Press, Champaign (2021)

Wasserman, H.: Power, meaning and geopolitics: ethics as an entry point for global communication studies. J. Commun. **68**, 441–451 (2018)

Xu, Y.-C.: Chinese state-owned enterprises in Africa: ambassadors or freebooters? J. Contemp. China **23**(89), 822–840 (2014)

Gu, J., Chuanhong, Z., Vaz, A., Mukwereza, L.: Chinese state capitalism? Rethinking the role of the state and business in Chinese development cooperation in Africa. World Dev. **81**(May), 24–34 (2016)

Calzati, S.: Decolonising 'data colonialism': Propositions for investigating the realpolitik of today's networked ecology. Television & New Media (2020). https://doi.org/10.1177/152747 6420957267

Mueller, M.: Sovereignty and cyberspace: Institutions and Internet governance (2019). https://www.intgovforum.org/multilingual/sites/default/files/webform/week13-cyberspacesovereignty.pdf

Yu, H., Goodnight, T.: How to think about cybersovereignty: the case of China. Chin. J. Commun. **13**(1), 8–26 (2020)

Gagliardone, I.: The Politics of Technology in Africa: Communication, Development, and Nation-Building in Ethiopia. Cambridge University Press, Cambridge (2016)

Kushwaha, N., Watson, B, Roguski, P.: Up in the air: Ensuring government data sovereignty in the cloud. In: 2020 12th International Conference on Cyber Conflict, Tallinn (2020)

Timmers, P.: Investment policy for digital sovereignty: From policy to action. In: Digital sovereignty: From narrative to policy?, pp. 25–33 (2023). https://eucyberdirect.eu/research/digital-sovereignty-narrative-policy

Hummel, P., Braun, M., Tretter, M., Dabrock, P.: Data sovereignty: a review. Big Data Soc. (2020). https://doi.org/10.1177/2053951720982012

Micheli, M., Ponti, M., Craglia, M., Berti Suman, A.: Emerging models of data governance in the age of datafication. Big Data Soc. **7**(2), 1–15 (2020)

Glasze, G., et al.: Contested spatialities of digital sovereignty. Geopolitics, 1–40 (2022)

Floridi, L.: Digital's cleaving power and its consequences. Philos. Technol. **30**, 123–129 (2017)

European Union. (2019). Policy and investment recommendation for trustworthy AI. https://ec.europa.eu/digital-single-market/en/news/policy-and-investment-recommendationstrustworthy-artificial-intelligence

Robles-Carrillo, M.: European 5G policy: Legal and geopolitical approach. In: Digital sovereignty: From narrative to policy? pp. 58–66 (2023). https://eucyberdirect.eu/research/digital-sovereignty-narrative-policy

Broeders, D., Kaminska, M.: EU digital sovereignty: when top-down meets bottom-up. In *Digital sovereignty: From narrative to policy?* pp. 9–17 (2023). https://eucyberdirect.eu/research/digital-sovereignty-narrative-policy

Voss, G., Pernot-Leplay, E.: China data flows and power in the era of Chinese big tech. Northwest. J. Int. Law Bus. (2023). https://doi.org/10.2139/ssrn.4393008

Brown, T.: Human rights in the smart city: regulating emerging technologies in city places. In: Reins, L. (ed.) Regulating New Technologies in Uncertain Times, pp. 47–65. Asser Press (2019)

Smuha, N.A.: Beyond the individual: governing AI's societal harm. Internet Pol. Rev. **10**(3) (2021). https://doi.org/10.14763/2021.3.1574

Taylor, L., Floridi, L., van der Sloot, B.: Introduction: a new perspective on privacy. In: Taylor, L., Floridi, L., van der Sloot, B. (eds.) Group Privacy. PSS, vol. 126, pp. 1–12. Springer, Cham (2017). https://doi.org/10.1007/978-3-319-46608-8_1

Viljoen, S.: A relational theory of data governance'. Yale Law J. **131**, 573 (2021)

Calzati, S.: Federated data as a commons: a third way to subject-centric and collective-centric approaches to data epistemology and politics. J. Inf. Commun. Ethics Soc. (2022). https://doi.org/10.1108/JICES-09-2021-0097

Taylor, L.: Public actors without public values: legitimacy, domination and the regulation of the technology sector. Philos. Technol. **34**(4), 897–922 (2021)

BMWi. GAIA-X: The European project kicks off the next phase. Germany Federal Ministry for Economic Affairs and Energy (2020). https://www.bmwk.de/Redaktion/EN/Publikationen/gaia-x-the-european-project-kicks-of-the-next-phase.pdf?__blob=publicationFile&v=13

Monti, G. EU competition law and digital sovereignty. In: Digital sovereignty: From narrative to policy? pp. 46–52. https://eucyberdirect.eu/research/digital-sovereignty-narrative-policy

Rochel, J.: Ethics in the GDPR: a blueprint for applied legal theory. Int. Data Priv. Law **11**(2), 209–223 (2021)

Zygmuntowski, J.J., Zoboli, L., Nemitz, P.: Embedding European values in data governance: a case for public data commons. Internet Policy Rev. **10**(3), 1–29 (2021)

Ostrom, E.: Governing the Commons: The Evolution of Institutions for Collective Action. Cambridge University Press, Cambridge (1990)

Dulong de Rosnay, M., Stalder, F.: Digital commons. Internet Pol. Rev. **9**(4), 1–22 (2020)

Morozov, E., Bria, F.: Rethinking the smart city: Democratizing urban technology. 5. City Series. Rosa Luxemburg Stiftung, New York (2018). https://rosalux.nyc/rethinking-the-smart-city-democratizing-urban-technology/

Bloom, G., Raymond, A., Tavernier, W., Siddarth, D., Motz, G., Dulong de Rosnay, M.: A practical framework for applying Ostrom's principles to data commons governance. https://foundation.mozilla.org/en/blog/a-practical-framework-for-applying-ostroms-principles-to-data-commons-governance/

Sanfilippo, M., Frischmann, B.: A proposal for principled decision-making: beyond design principles. In: Frischmann, B., Madison, M., Sanfilippo, M. (eds.) Governing Smart Cities as Knowledge Commons, pp. 295–308. Cambridge University Press, Cambridge (2023)

de Angelis, M.: Omnia Sunt Communia: On the Commons and the Transformation to Postcapitalism. Bloomsbury Publishing, London (2017)

Author Index

© IFIP International Federation for Information Processing 2023
Published by Springer Nature Switzerland AG 2023
N. Edelmann et al. (Eds.): ePart 2023, LNCS 14153, p. 167, 2023.
https://doi.org/10.1007/978-3-031-41617-0

Printed in the United States
by Baker & Taylor Publisher Services